Harvey's Heart

The Discovery of Blood Circulation

Andrew Gregory

Series editor: Jon Turney

ICON BOOKS UK

TOTEM BOOKS USA

Published in the UK in 2001
by Icon Books Ltd., Grange Road,
Duxford, Cambridge CB2 4QF
E-mail: info@iconbooks.co.uk
www.iconbooks.co.uk

Published in the USA in 2001
by Totem Books
Inquiries to: Icon Books Ltd.,
Grange Road, Duxford,
Cambridge CB2 4QF, UK

Sold in the UK, Europe, South Africa
and Asia by Faber and Faber Ltd.,
3 Queen Square, London WC1N 3AU
or their agents

Distributed to the trade in the USA
by National Book Network Inc.,
4720 Boston Way, Lanham,
Maryland 20706

Distributed in the UK, Europe,
South Africa and Asia by
Macmillan Distribution Ltd.,
Houndmills, Basingstoke RG21 6XS

Distributed in Canada by
Penguin Books Canada,
10 Alcorn Avenue, Suite 300,
Toronto, Ontario M4V 3B2

Published in Australia in 2001
by Allen & Unwin Pty. Ltd.,
PO Box 8500, 83 Alexander Street,
Crows Nest, NSW 2065

ISBN 1 84046 248 5

Series editor: Jon Turney

Originating editor: Simon Flynn

Typesetting by Hands Fotoset

Printed and bound in the UK by
Cox & Wyman Ltd., Reading

Contents

Acknowledgements and Dedication   iv

Introduction   1

1 The Background to Harvey's Discovery   8
*Galen 9; Galen on the Heart and Blood 12; Harvey 22;
The Renaissance 24; Vesalius 26; Fabricius 31;
Renaissance Art 34; Renaissance Art and Anatomy 35*

2 Harvey's Discovery of the Circulation   47
*De Motu Cordis 52; The Circulation 54; The Venous
Valves 57; The Ligature Experiments 59; The Flow-Rate
Experiment 66; The Motion of the Heart 69; The Pulse
and the Heart Valves 73*

3 The Nature of Harvey's Discovery   79
*The Scientific Revolution 80; Harvey and Aristotle 86;
Harvey and Quantification 88; Harvey and Mechanical
Models 92; Harvey and the Natural Magic Tradition 94;
Harvey and Witchcraft 105; Harvey and the Scientific
Revolution 108*

4 The Reception of Harvey's Discovery   111
*Opposition to Harvey 115; The Theory of 'Ebullition' 119;
What is Actually Seen? 122; Other Objections 125; The
Dispute with Descartes 133; Harvey's Later Work 139*

Conclusion   142

Further Reading   147

Glossary   150

Acknowledgements

I would like to thank the editors, Jon Turney and Simon Flynn, for their patience and efficiency and for their comments on the manuscript. I would also like to thank Ms Sheelagh Doherty RGN, RSCN, RM for her support, her comments and for checking the manuscript for medical accuracy. Without their help this would have been a less interesting and less accurate book.

Dedication

For Sheelagh, *with love*

Introduction

At some time around 1618, William Harvey, an English physician, made a remarkable discovery that was to revolutionise thinking about the human body. He found that the blood circulated around the body, and did so rapidly. He also discovered a great deal about the motion and function of the heart, believing it to pump blood around the body. Today, we take these to be basic and evident facts. So why were these discoveries so remarkable and so momentous?

In the era of open-heart surgery, arterial bypasses and blood transfusions, we believe we understand the motion of the heart and the blood very well. Doctors are familiar with the sight and feel of the human heart pumping away, and TV documentaries bring these images into our homes. Such easy access to the human heart has not always been available, however. Prior to anaesthetics, blood transfusions and antiseptics, to open the chest and expose the heart meant certain death for the patient. Knowledge came only through animal vivisection and human post-mortem dissection, and both methods had significant drawbacks. One can discern the structure of the heart relatively easily from dissection, but not its motion and function. Vivisection was no great help, as the motion of the heart is swift and in a distressed subject likely to be either rapid or disorganised. So too, as soon as one opens a living heart to see its internal workings, the subject is likely to die. The motion of the heart is also complex. How many of us could say with certainty exactly how the heart beats? Where does the contraction of the heart begin and where does it finish, or does it all contract at once? What function do the heart valves perform, and when do they open and shut? The heart rises and falls in the chest as it beats. Which of these is muscular contraction and which relaxation? At which point does the heart have its largest and smallest volume? The next time you see a documentary showing open-heart surgery, try discerning the basic motion of the heart yourself. Initially, it is by no means easy, even if you know what you are looking for. For those who know a good deal about the motion of the heart, remember it is one thing to look when you know what you are looking for, quite another to work out what is happening with no prior information, or even worse, in Harvey's case, with incorrect prior information. The heartbeat has a considerable cultural significance as a sign of life, both now and in the seventeenth century. But how, exactly, does the heart beat and what is its purpose? That was a key question that Harvey hoped to solve.

How the heart is connected to the major blood vessels that bring blood to it and take blood from it is also reasonably evident from dissection. It is not in the least bit evident, however, how those vessels and their tributaries relate to each other in parts of the body remote from the heart. Do they form open-ended systems, as was commonly thought prior to Harvey? Or are there several closed systems? Or do they join in one closed system? In an age when the capillaries, the fine blood vessels that join the arteries to the veins, were unobserved because of their minuteness, this was not an easy question. Nor was it clear how quickly the blood flowed. Opening blood vessels is no great help here. Certainly we bleed quickly if cut in certain places, but if you make a hole in a water pipe when water is not flowing, it will still leak rapidly. Harvey's discovery of the rapid circulation of the blood and the proper motion and function of the heart was remarkable relative to the knowledge, technology and methods available in the seventeenth century. Harvey's discoveries had to be inferred, not merely observed. He had to create a good number of ingenious experiments and well-argued inferences without ever having seen the internal workings of human beings.

Harvey's discoveries were also quite contrary to accepted opinion, which had stood for some 1,500 years. Everyone but Harvey thought blood was generated in the liver and was gradually consumed by the body, giving it nutrition. Blood flowed very slowly away from the liver to the other parts of the body, but not back to the liver. This was the view of Galen, greatest of the ancient anatomists, and accepted without question down to Harvey's time. Although Galen's views may seem odd to the modern eye, in the context of their time, they offered a plausible and comprehensive account of the human body. Galen's account of the heart and arteries was attuned to a very slow flow of the blood. He could easily explain rapid bleeding without a rapid circulation. Galen's views were well entrenched, and were defended vigorously by intelligent, experienced men. Anatomists in sixteenth-century Italy began to examine the human body with the aim of improving on Galen, but no one suspected that he was radically wrong regarding both the motion of the heart and the blood. Harvey had two significant obstacles to overcome. There was the difficulty of conceiving not only of the circulation of the blood, but also of an account of the heart and arteries that would suit a rapid circulation theory, against 1,500 years of tradition. Harvey had to amass sufficient evidence and arguments in favour of his new ideas to win over some highly sceptical Galenic opponents, and that, as we shall see, was no easy matter.

Their doubts were increased because Harvey's ideas affected our whole conception of how the body works. If Harvey was right about the cardio-vascular system, then Galen was wrong, not merely about the heart and blood, but also about digestion, nutrition, respiration and aspects of organ function. Galen's physiology depended to a large extent on principles of attraction. Parts of the body (including the heart and the arteries) were thought to attract what they required to themselves. Harvey's work showed this was quite wrong for the heart and blood, as the heart forces blood around the body. This prompted a complete re-evaluation of the attraction principle. A significant part of medical therapy in Galen's system was based on blood letting. Blood was drawn from specific parts of the body, determined by Galen's conception of the blood system and the position of the organ thought to be affected, in order to treat a disease. If Galen was wrong on the blood system, however, this entire mode of treatment needed to be re-thought. The deeper significance of Harvey's discovery, then, was that it helped instigate a complete rethink of the way that the body works and how it should be treated. Ultimately, Galen's anatomy and physiology were not in need of improvement, as the Renaissance anatomists had attempted, but complete rejection.

What led Harvey to such a momentous discovery? What were the experiments and arguments that Harvey used to support his theory? Who and what did he have to struggle against in order to get this remarkable and controversial new idea accepted? These are some of the questions that this book will try to answer. Firstly though, we need to look briefly at the three ancient Greeks Harvey most admired, Hippocrates, Aristotle and Galen. We also need to look at the nature of the Renaissance, and at the advances in anatomy made by Vesalius and Fabricius, in order to understand the state of anatomy, physiology and medicine as Harvey found it (anatomy studies the structure of the body, while physiology studies its function).

One final comment before we begin. Harvey carried out many experiments, often with fatal consequences, on live animals. It is necessary to describe these, in order to understand Harvey's arguments in favour of the circulation of the blood. This should not be taken to imply that I condone such experiments. As a matter of historical fact, vivisection and the use of animals in research were taken for granted in the seventeenth century and were not matters of debate. The seventeenth century simply did not share, and indeed had no conception of, our twentieth- and twenty-first-century attitudes towards the rights of animals.

1 The Background to Harvey's Discovery

The beginnings of the systematic study of the human body can be traced back to the ancient Greeks and a collection of books known as the Hippocratic corpus. These were written by Hippocrates (460–370 BC) and his followers around the fifth to the third century BC. The Babylonians, the Egyptians and other societies prior to the Greeks had a rudimentary and practical knowledge of anatomy and healing techniques. It was the Greeks, however, who began to systematise this knowledge. They introduced proper scientific method, and emphasised that diseases were due solely to the body, and not the intervention of the gods. The Hippocratics knew a fair amount about the structure of the heart, and recognised it as a muscle. They were less sound, though, on its function and movement. That was a recurring theme up until the work of Harvey. It was relatively easy to dissect a dead heart and discern its structure. It was something else entirely to work out precisely how the living heart moves, what it does, and how it relates to the blood.

Aristotle (384–322 BC) was significant both as a philosopher and an anatomist. One of the great names in the history of philosophy, Aristotle was also deeply interested in human and animal anatomy, and was effectively the founder of biology. He emphasised the importance of the cardiovascular system in the higher animals. He also did important work on describing the blood system, especially the deeper lying vessels, about which little was known at the time. He failed, however, to distinguish between arteries and veins, and believed that the heart, not the brain, was the organ of sensation.

Galen

Of the ancients, the Roman physician and scholar Galen (130–200 AD) is by far the most important for us. It was his views that Harvey had to struggle to replace. Galen was an admirer and follower of Hippocrates. He made his own original contributions to ancient medicine and anatomy, while systematising existing knowledge in these areas. He pioneered the use of the pulse in diagnosis. At various times Galen was physician to the Roman army, gladiators and Roman emperors. Whatever his experience of the human body in the amphi-theatre, Galen faced one major problem. Access to human bodies for dissection in his time was virtually impossible, due to social taboos concerning the dead body. So we can find him saying that:

It is possible to see something of human bones. I have done so often on the breaking open of a tomb or grave. Thus once a river, inundating a recently made grave, broke it up and swept away the body of the dead man. The flesh had putrefied, but the bones still held together in their proper relations.

There was a brief period in the ancient world when dissection and even vivisection of humans was possible. Celsus, a Roman physician of the first century AD, tells us that:

Since pains and various kinds of disease arise in the internal parts, they hold that no-one who is ignorant of these parts can apply remedies to them. Therefore it is necessary to cut open the bodies of dead men and to examine their viscera and intestines. Herophilus and Erasistratus proceeded in by far the best way: they cut open living men – criminals they obtained out of prison – and they observed, while their subjects still breathed, parts that nature had previously hidden.

The debate about the merits of dissection and vivisection, and how much of human anatomy can be learnt by studying animals, recurred in the sixteenth and seventeenth centuries. Greek medicine, anatomy and physiology were all held back by a lack of an adequate number of human bodies to dissect. The standard ancient theory of disease, which Galen helped to clarify and codify, involved the four humours of the body: blood, phlegm, yellow bile and black bile. When these were in balance, the patient was healthy. Disease was an imbalance of the humours. Here was the justification of the practice of leechcraft and blood letting. If the diagnosis of your illness was that your humours were out of balance due to an excess of blood, then the clear path to health was to remove some of that blood.

Galen had some distinctive and very influential ideas on the motion of the heart and the blood. These ideas lasted for well over 1,500 years and found defenders even in the seventeenth century. Why was this so? The motion of the heart and the blood is not easy to investigate. No one can see the circulation of the blood, as such. The circulation is a theory that we adopt because we believe it is the best one to explain the facts as we know them. Galen's theory could cover the facts as known to him surprisingly well, and it fitted well with his other anatomical and physiological theories. Galen's theories require some close attention.

Galen on the Heart and Blood

Galen believed there were two quite separate blood systems in the human body, operating simultaneously. Blood did not circulate, in the sense of flowing around a closed system, but was gradually produced by the liver and consumed by the body.

One system, based on the liver, distributed nutrition to the body. Food was digested in the intestines and turned into a substance called chyle. Nutrition was carried to the liver via the portal vein. The liver then generated nutritive blood, which was

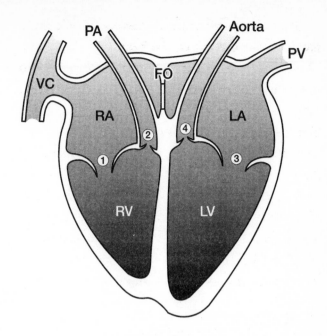

Figure 1: The structure of the heart.

Key:

VC	Vena Cava
RA	Right Atrium
RV	Right Ventricle
PA	Pulmonary Artery
FO	Foramen Ovale (open in foetus but closed after birth)
LV	Left Ventricle
LA	Left Atrium
PV	Pulmonary Vein
1	Tricuspid valve
2	Pulmonary valve
3	Bicuspid (mitral) valve
4	Aortic valve

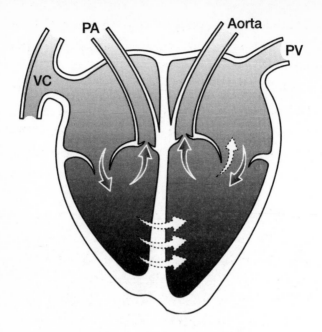

Figure 2: The flow of blood in the heart according to Galen.

distributed to the rest of the body via the veins. The liver was believed to produce blood because it looked like congealed blood; its proximity to the stomach and intestines led to the belief that it was the seat of nutrition. Some of this nutritive blood (but not all) entered the right side of the heart and was passed on to the lungs. The nutritive blood was gradually consumed by the body in order to give it nutrition. This was an open system. Blood was generated in the liver, passed to the rest of the body and consumed there, giving it nutrition, but this blood did not return to the liver.

The body's second blood system, according to Galen, had the function of distributing a vital spirit to the body. Air was drawn into the lungs and this vital spirit (known to the Greeks as *pneuma*) became mixed with the blood. This vivified blood then

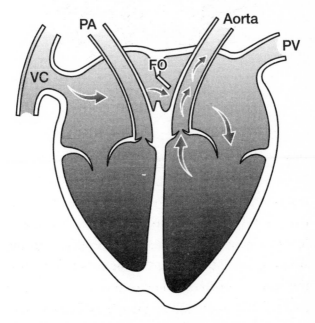

Figure 3: Foetal blood-flow. In the foetus, there is no blood-flow through the lungs, and the foramen ovale is open. When the baby takes its first breath, blood begins to flow through the lungs and the foramen ovale closes.

passed to the left side of the heart and was distributed to the rest of the body via the arteries. Again, this was an open system. Blood was vivified in the lungs, passed to the rest of the body and consumed. In both systems, blood moved around the body, but was thought to do so slowly.

There were supposed to be two connections between these two blood systems. As the body was believed to consume vivified blood as well as nutrified blood, the vivified blood would gradually deplete unless it were replenished. It was thought that a small amount of nutrified blood passed across the muscular central wall of the heart, the septum, and so replenished the vivified blood. Blood does not pass through the septum in healthy adults, but Galen had two reasons for believing that it did. He knew that in the foetus blood does pass through an opening in the septum. In fact the septum closes at birth, when the lungs begin to function. When this closure fails to take place fully, babies are said to have a 'hole in the heart', which can now be corrected by surgery.

Galen also believed that small indentations in the septum were the widest part of openings that became invisibly small but allowed the passage of blood. Whether or not there were pores in the septum was a notorious controversy in anatomy up to the acceptance of Harvey's theories. Galen also believed (quite correctly) that there were minute connecting passages between arteries and veins, called anastomases. However, he did not believe them to be great in number, nor did any significant amount of blood flow in them. The two systems were then effectively separate, with a small flow of nutritive blood replenishing the vivified blood consumed by the body.

Why might an intelligent and well-informed person have been led to believe this? Understanding Galen's view helps us understand the magnitude of the problem that Harvey faced. There are in fact two different types of blood, oxygenated and deoxygenated, and these have noticeably different colours. Oxygenated blood is bright scarlet, while deoxygenated blood is purple. When blood passes through the lungs, it passes through a fine network of capillaries around small sacs, known as alveoli. Oxygen from the air in the alveoli is exchanged for carbon dioxide, a waste product from the use of oxygen in the body. These small sacs give the lungs a much greater surface area and so allow for a greater exchange of these gases.

When this exchange of oxygen for carbon dioxide takes place, the deoxygenated blood becomes oxygenated and changes colour from purple to scarlet. The oxygenated blood from the lungs returns to the left side of the heart and is distributed via the arteries and then capillaries to the rest of the body. As the organs and the muscles use oxygen, the blood becomes deoxygenated and changes back from scarlet to purple. Blood returns to the heart via the capillaries and the veins, where it is then pumped out into the lungs, and again passes through the lung capillaries exchanging carbon dioxide for oxygen.

We know this modern account of how the two types of blood can constantly be changing into each other, but Galen did not and nor did Harvey. The two different colours seemed very strong evidence that there are two quite different sorts of blood, and two systems that contain them. Another important fact is that there are two different types of blood vessel. Arteries have noticeably thicker walls, which are more elastic than those of the veins. What is more, arteries carry a pulse, while veins do not. The pulse is caused by the beating of the heart, which when it expels blood creates a pressure wave in the arteries. For Galen, the division of types of blood vessel was simple, and according to the type of blood they carried. One type of vessel carried one type of blood. So the veins carried nutritive, purple blood and the arteries carried vivified, scarlet blood. The modern view, originating with Harvey, is that arteries carry blood away from the heart, while veins carry blood to the heart.

Galen also had a distinctive view of the heartbeat and pulse. He believed the heart's active phase was when it expanded, attracting blood into itself. The heart only contracted when it relaxed, and so for Galen the heart did not expel blood with any great force. Actually the reverse is the case, and the heart contracts with some force to expel blood. Since for Galen the active phase of the heart, expansion, was not in time with the pulse, Galen believed that the arteries pulsed of their own accord, expanding and then relaxing. This pulse attracted blood from the heart. This is not such an odd idea. Galen was well aware that the intestines expand and contract on their own and so pass food through the digestive system or, as he would put it, attract nutrition. With this view of the heart and pulse, one might well believe that no great amount of blood was passed by the heart, only that amount which the body consumed for its nutrition. Attraction played a critical role in Galen's physiology. Parts of the body attracted blood or nutrition to themselves, as they required it, and did not have it pumped or forced to them.

The heart does not change greatly in volume externally during beating, even though it changes in shape. Which motions of the heart showed expansion and contraction in volume, and which of them was action or relaxation, was hotly disputed up to the acceptance of Harvey's views. As late as the 1640s, some distinguished anatomists held a view similar to Galen's in opposition to Harvey.

Galen's theory could account for many of the basic phenomena associated with the blood system. While the blood did not move rapidly according to Galen, it would be under some pressure and so certain wounds would bleed rapidly. Arteries when cut in fact spurt blood in quite a spectacular fashion, but arteries pulsed of their own accord for Galen, and so spurted when cut. Galen's theory also gave a plausible account of the assimilation of food, the production of blood, the distribution of nourishment to the body, the heart beat and pulse, and the production and distribution of heat around the body. This theory was not a negligible opponent, either in the ancient world or the seventeenth century.

One can see, then, the attraction of Galen's theory, especially when one considers that there were serious difficulties in conceiving of a closed circulation of the blood. If there was only one system, how did blood get from arteries to veins? Remember that capillaries were quite unknown at this stage, and require a microscope to be visible: they were only discovered in 1660 by Marcel Malpighi. If there was only one system, how could there be two different types of blood in it? There was a need to conceive of a way in which one type of blood could be converted into the other, and back again. There was also the need to conceive of where this happened in the body to be able to give a reasonable account of variations in blood colour. For Harvey, the blood not only circulated, but it did so very rapidly around the body. This provided further problems. The faster the rate of flow, the greater the flow through the capillaries. So there must be many capillaries, and they must carry a considerable quantity of blood very quickly, yet these capillaries could not be seen. There must be a great number of unseen capillaries in the lungs as well, as in the circulation theory the lungs have all of the blood circulating through them.

Harvey

Harvey was in many ways a man of the Renaissance rather than the later seventeenth-century scientific revolution. It is important to understand the nature of the Renaissance and especially Renaissance anatomy. There were two developments in the Renaissance that were very important for anatomy. Firstly, there was a new and much more positive attitude towards discovery and learning. Secondly, there were important developments in art, which inspired advances in science in general, and anatomy in particular.

After the successes of Greek science and medicine, with the Hippocratics and Galen to the fore, came the decline and fall of the western part of the Roman Empire around 500 AD, and the onset of the Dark Ages (500–1000 AD) and the Middle Ages (1000–1400 AD). With the sack of Rome, the great majority of the ancient Greek texts were lost to the West, which underwent a period of relative intellectual stagnation. The later Middle Ages saw a revival, to some extent, and two quite distinct traditions of learning. There was the scholarly tradition, which based itself in the monasteries and universities, and was almost entirely theoretical.

The eastern Roman Empire had survived longer, and the Islamic culture that succeeded it had preserved the ancient Greek texts, which were gradually transmitted to the West. The Islamic culture did more than merely preserve the works of the ancient Greeks. It absorbed them and made significant advances on them, especially in medicine, in which Avicenna and Rhazes (both *c*.1000 AD) were significant contributors. However, the scholarly tradition saw its main task as the recovery and interpretation of the ancient Greek texts and their harmonisation with Christian theology. It was interested in the ordering and clarification of what was already known, especially in relation to the ancient texts, rather than the discovery of anything new. Only where theology and philosophy touched on the natural world did it deal with what we would call matters of science and, when it did so, it used discussion and debate rather than experiment.

The other stream of learning was the craft tradition, centred around guilds, such as the weavers, potters or metalworkers. Knowledge of their craft practices was passed down through apprenticeships in a relatively secretive manner. This knowledge was almost entirely practical in its orientation and, although some slow progress in technology was made, there was little attempt to form any general laws or theories. The scholastic and the craft traditions had very little to say to each other, being based on entirely different strata of society and having radically different aims.

The Renaissance

The Renaissance, literally 'the rebirth', began around 1400 in Italy, initially as an artistic movement. It took the task of art and science to be not merely to recover the ancient texts, but to recreate a perceived golden age of ancient Greece and develop it further. Many things were done for the first time since antiquity, such as casting life-size bronze statues. Renaissance achievements in engineering and art soon outstripped the ancients. The Renaissance generated a great new optimism towards learning and discovery and what could be done with it. Renaissance thinkers looked back and revered antiquity and looked forward to creating something new and better at the same time. The Renaissance was quite distinct from the Middle Ages, which lacked this sense of optimism and progress. It was also distinct from the scientific revolution, which attempted to discard the ancients altogether and replace them with something new. The Renaissance also saw the rise of the polymath, the person interested in learning in general, not merely in one narrow, specialised discipline.

These differences in attitude between the Middle Ages and the Renaissance can be seen in the way that dissections of the human body were conducted. There was a rigid hierarchy and division of labour at a dissection, between the lecturing professor, the ostensor (a man who pointed out the parts being discussed) and the barber surgeon who cut up the corpse. At this stage, surgery was considered a manual, low-class job which was carried out by barbers rather than doctors (this is why barbers' poles used to have red and white stripes). The professor would read from a text, but would take no part in the actual dissection of the body. The ostensor would point to the parts being discussed by the professor, while the barber surgeon would dissect the body but would not speak during the dissection. The critical difference between medieval and Renaissance dissections can be put like this. Medieval dissections were demonstrations of what was already known, rather than investigations intended to discover anything new. Renaissance anatomists, while still depending heavily on the works of the ancients, began to attempt to correct and improve on them.

There is something of a myth that anatomy, prior to and during the Renaissance, was held back by a lack of bodies to dissect. This is, at best, only partly true. What really held anatomy back was the lack of an inquisitive attitude. When dissections were held as demonstrations, there was no great need for more bodies. Doubtless it would have been nice to have had more, but no one agitated vigorously to get them. When dissections began to be held for the purposes of investigation and discovery, more bodies were required, people agitated to get them and, to a certain extent, they were successful. While it was never easy or straightforward to get hold of corpses, Colombo claims to have dissected fourteen in one year and, in 1539, when Contarini, a Paduan judge, became interested in Vesalius' work, he made the bodies of executed criminals available to him. Harvey also dissected miscreants who had been hanged.

Vesalius

Andreas Vesalius (1514–1564) was a key figure in

Renaissance anatomy and was enormously important in breaking down old attitudes. Vesalius began as an ardent supporter of Galen, and early in his life wrote a work comparing Islamic and Galenic theories, favouring Galen's theories. From this traditional beginning, Vesalius became more radical. He broke down the division of labour at dissections and not only lectured but did his own dissecting. He insisted this was the proper way to do anatomy. Vesalius was critical of the old hierarchy, and said that:

The more fashionable doctors, first in Italy, in imitation of the old Romans, despising the work of the hand, began to delegate to slaves the manual attentions they judged needful for their patients, and themselves merely stood over them like architects. Then when all the rest who practised the true art of healing gradually declined the unpleasant duties of their profession, without however abating any of their claim to money, or to honour, they fell quickly away from the standard of the doctors of old. Methods of cooking, and all the preparation of food for the sick, they left to nurses; compounding of drugs they left to apothecaries;

manual operations to barbers . . . When the whole conduct of manual operation was entrusted to barbers, not only did physicians lose the true knowledge of the viscera, but the practice of dissection died out, doubtless for the reason that the doctors did not attempt to operate, while those to whom the manual skill was resigned were too ignorant to read the writings of the teachers of anatomy.

It is not good enough to do one dissection, he says, one must do many. Moreover, he did some of his own anatomical drawings and commissioned good artists to do others. Anatomical texts prior to Vesalius were very poorly illustrated. Here we see the importance of the Renaissance polymath, the person interested in all aspects of learning and its application and not in just one narrow, theoretical speciality. The notion of the polymath undoubtedly helped the cross-fertilisation of ideas between art and science in the Renaissance. Vesalius' major work was called *De Humani Corporis Fabrica* (On the Fabric of the Human Body, *De Fabrica* for short), and was published in 1543. *De Fabrica* did not follow the traditional pattern for anatomical texts of looking at the abdomen first, but began with the bones, as these, Vesalius believed, were the foundation of the human body. The reason why old texts began with the abdomen was that there was limited time for dissection before the body began to putrefy. The abdomen was usually the first part to go off, so it was dealt with first. It was also thought to be the lowest part in a hierarchy of abdomen, chest and head. Vesalius' approach was undoubtedly more logical anatomically, though he was helped by the fact that he had more corpses to work with.

The most radical aspect of Vesalius' work was that he began to point out anatomical errors in the work by Galen. Galen, unable to dissect anything like a sufficient number of human beings, had dissected a great number of the higher mammals. The basic problem with Galen's anatomy was that he assumed too great a likeness between these higher mammals and humans. For instance, Galen's description of the muscles of the human hand is very good, but not quite right. It is, however, an excellent description of the hand of a Barbary ape. Galen believed that a fine network of arteries at the base of the brain, called the *rete mirabile*, could be found in humans. In fact it cannot, but it can be found in sheep. There are many more such examples. The existence of the *rete mirabile* in humans had been questioned before, but was now specifically denied by Vesalius. Vesalius declared that the only way to learn about human anatomy was to dissect humans and, in a radical move in 1540, dissected human and ape skeletons to show that some of Galen's descriptions were of apes and not humans.

Vesalius' pioneering work created a Renaissance industry of improving anatomy by removing some of Galen's errors. However, this did not produce a major break from Galen's basic ideas about the nature and function of the human body. Nor did it produce any really new thinking on the heart or the blood vessels, although there was a debate about the passage of blood across the septum. Vesalius stated:

Not long ago I would not have dared diverge a hair's breadth from Galen's opinion. But the septum is as thick, dense, and compact as the rest of the heart. I do not see therefore, how even the smallest particle can be transferred from the right to the left ventricle through it.

There was great respect for the opinions of ancient thinkers in the Renaissance and people did not go against their views willingly, as we can see here. When they did, the intention was to improve upon the works of the ancients, rather than to replace them.

Fabricius

Another important figure in Renaissance anatomy was Girolamo Fabrici (1533–1619), known as Fabricius. He was instrumental in building and inaugurating the first permanent dissection theatre in 1595, which is still standing. Previously dissections had taken place only in a two-week period in each year and only temporary theatres were erected. His most important work, for our purposes, was *De Venarum Ostiolis* (On the Little Doors of the Veins), published in 1603, though he published much other work of merit as well. This was an abundantly and beautifully illustrated book on the valves that can be found in the veins. Fabricius did not realise the proper function of these valves or their significance for the flow of the blood, though Harvey did. Vesalius was professor of anatomy at Padua while Harvey studied there, and his methods and programme had a considerable effect on Harvey.

One significant advance made during the Renaissance was the discovery of what was known as the lesser circulation – or the 'pulmonary transit' (see Figure 4). Here blood was thought to pass from the right side of the heart through the pulmonary artery to the lungs. It then passed through the lungs and returned via the pulmonary vein to the left side of the heart. Blood was still thought to be produced by the liver and consumed by the body, so here we have one open-ended system instead of Galen's two, or the closed system of the full circulation. Three people put this idea forward, as far as we know independently of each other. Firstly, in the thirteenth century, working within the Islamic tradition, there was Ibn al-Nafis. In sixteenth-century Italy, there was Realdo Colombo (1510–1559) and Michael Servetus (1511–1553). It is unlikely that Harvey knew of Ibn al-Nafis and unlikely too that he had heard of Servetus' theory. However, Harvey quoted the work of Colombo several times and, although he did not refer directly to the theory of the lesser circulation, it is quite possible that he was aware of Colombo's version of it. Colombo also began to question whether Galen's account of the motion of the heart was correct, and Harvey seems to have

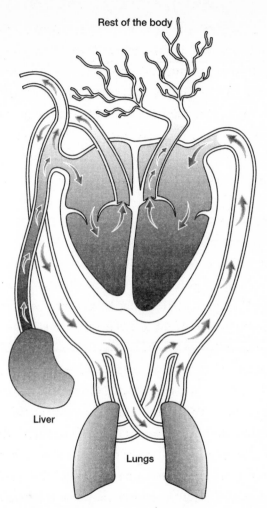

Rest of the body

Liver

Lungs

Figure 4: Pulmonary/lesser circulation. Blood is thought to flow from the heart, through the lungs and back to the heart again. This is not the full circulation, however. Blood is still generated by the liver and consumed by the body.

been aware of this as well. Colombo also pointed out some errors in the work of Vesalius, which led to some friction between the two of them, Vesalius claiming that Colombo was a scoundrel and that he had taught Colombo all he knew.

Renaissance Art

The art of the Renaissance also made an important contribution to science and anatomy. The Renaissance saw a new realism in art, in contrast to the more figurative style that preceded it. In particular, there was the development of mathematical theories of proportion and perspective that gave pictures much greater depth and focus, in sharp contrast to the rather 'flat' paintings of the Middle Ages. The use of mathematics was critical to the way in which Renaissance art depicted the world. This idea that mathematics is critical for an accurate description of the world was also important for the development of Renaissance science. Was the world intrinsically mathematical, something that could be accurately described in numbers and geometrical shapes? Tradition was against this, as the dominant philosophy of the time, scholasticism, denied that the world was like this and saw no important role for mathematics in science. However, there were those who had a different vision of the world. As Galileo famously wrote:

> *Philosophy is written in that vast book which stands forever open before our eyes, I mean the universe; but it cannot be read until we have learnt the language and become familiar with the characters in which it is written. It is written in mathematical language, and the letters are triangles, circles and other geometrical figures, without which means it is humanly impossible to comprehend a single word.*

While we would not put this in quite the same terms today, this basic idea that mathematics is the key tool for science, and especially the physical sciences, has proved to be a most fruitful one.

Renaissance Art and Anatomy

Critical for our story, however, is the way in which this enthusiasm for the accurate depiction of nature feeds into anatomy. Prior to the Renaissance, anatomy textbooks had either been very poorly illustrated or not illustrated at all. A very important piece of work here is Leonardo da Vinci's proportions of the human body, commonly known as the *Vitruvian Man*. Leonardo made other anatomical sketches, but this one is important in suggesting that the human body can be depicted most accurately in a geometrical manner.

We can see a radical change in the textbooks of the anatomists of the Renaissance, beginning with Vesalius. *De Fabrica* is a lavishly illustrated work, full of beautiful and accurate depictions of various parts of the human body. Vesalius probably employed students of Titian to do the diagrams, and he was careful to hire excellent woodcutters to generate the woodcuts from which the diagrams in *De Fabrica* were printed.

We can see these two themes of increased interest in new learning and the new realism in art in comparisons between depictions of dissections and diagrams from medical texts. Firstly we have woodcuts of dissections from 1493, and for comparison the frontispiece of Vesalius' *De Fabrica* from 1543. The differences, even at first glance, are quite startling. The woodcuts seem crude and lifeless in many ways, lacking in proportion and perspective, without strong focal points and generally having a much more medieval feel.

Figure 5a: A woodcut of an anatomical dissection taken
from *Anathomia*, ed. M. P. von Mellerichstadt, Leipzig:
Landsberg, *c.* 1493. (Source: Wellcome Library, London)

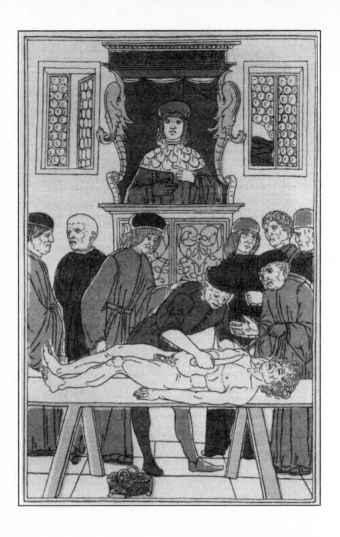

Figure 5b: A woodcut of an anatomical dissection, overseen by a seated professor, taken from the *Fasciculo di medicina*, J. Ketham, Venice 1493. (Source: Wellcome Library, London)

Figure 6: The woodcut title page of *De Humani Corporis Fabrica* (1543) by Andreas Vesalius. (Source: Wellcome Library, London)

Vesalius' frontispiece is a completely different matter. Here we have a magnificent example of the new use of proportion and perspective, and there are many typical Renaissance devices to draw your eye to the central focal point. So too, the characters have that glorious Renaissance animation to them, of people interested in a quest for knowledge.

In the 1493 woodcuts, we can clearly see the division of labour between the professor reading from a text and the barber surgeon actually performing the dissection. Dissection also often employed the services of an ostensor (see Figure 5b), whose job was to point out the parts of the body being referred to by the professor as the barber surgeon laid them bare.

Vesalius himself is the central character in his own frontispiece, and he has broken down this division and is both dissecting and teaching. As he did some of his own drawings as well, he has good claims to be considered a Renaissance polymath. Look at the spectators too. In Figure 5b, no one seems particularly interested in the dissection. Perhaps the characters talking to each other are debating points of anatomy, perhaps they are just gossiping, but there is no great animation about them. Compare the Vesalius frontispiece, in which everyone is clamouring to see and to learn. That seems representative of the new attitude of the Renaissance.

The three large characters in ancient dress in the foreground are supposed to be ancient Greek anatomists, who are also interested in the new knowledge. Vesalius does not want to break with the ancients, so they are given heroic poses and an important place, but they too wish to learn new things about the body. Vesalius, like so many in the Renaissance, looks back to antiquity whilst attempting to push its achievements further forward.

The skeleton may be there to emphasise the importance of the bones as the foundation of the body, an important matter for Vesalius in *De Fabrica*, although it may simply be a typical Renaissance motif as a reminder of man's mortality.

Let us now compare a diagram from the pre-Vesalius period (Figure 7) with some of those we can find in *De Fabrica* (Figures 8 and 9). We can see here a rather figurative representation of the human body, and this would be one of a very few illustrations. It gives some idea of the relation of the parts of the body, but not an accurate depiction. If

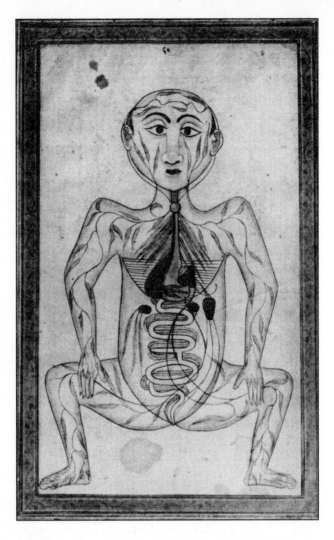

Figure 7: Anatomical print from the pre-Vesalius period.
(Source: Wellcome Library, London)

Figure 8: 'Quinta musculorum', from *De Humani Corporis Fabrica* (1543) by Andreas Vesalius. (Source: Wellcome Library, London)

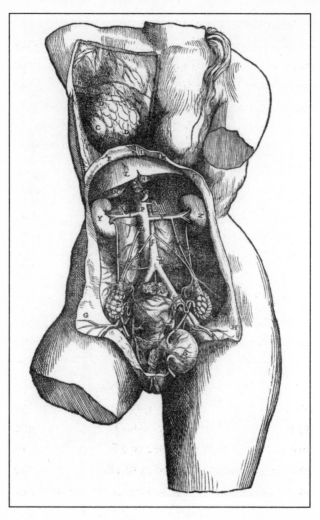

Figure 9: Female torso from *De Humani Corporis Fabrica* (1543) by Andreas Vesalius. (Source: Wellcome Library, London)

we compare this with the depictions of parts of the human body in Vesalius' book (and he has very many of these depictions), the differences again are striking. We can see for ourselves the fruits of the new realism and the refined ideas about proportion and perspective in the Renaissance, and how these have fed into the science of anatomy.

This, then, was the situation that Harvey stepped into. The main figure in anatomy was still that of Galen. His basic ideas on the nature of the human body had survived, even if such Renaissance pioneers such as Vesalius and Fabricius had removed some of the minor errors. Ibn al-Nafis, Servetus and Colombo had all independently suggested the lesser circulation, of heart to lungs and back to the heart again, and it is quite possible that Harvey was aware of Colombo's theory. The lesser circulation was by no means universally accepted, however, and the question of pores in the septum was still under debate. There were some serious problems for the idea of a full circulation, of how the two types of blood could be present in the one system and how the two systems supposed by Galen could be connected together. The Renaissance's renewed enthusiasm for learning and the new realism in art gave new vitality to anatomy and allowed some limited criticism of the ancients. Galen's theories of the heart and blood were still pretty much intact, however, and it was these that Harvey had to contend against.

2   Harvey's Discovery of the Circulation

William Harvey was born on 1 April 1578 at Folkestone in Kent, and died on 3 June 1657. His father was a farmer and landowner, and he was one of a family of seven brothers and two sisters. He went to King's School, Canterbury, and then on to study arts and medicine at Gonville and Caius College, Cambridge from 1593 to 1599. He then went to Padua, the leading European medical school, and became a student of Fabricius, gaining his doctorate on 25 April 1602. He returned to England and took up practice in medicine. In 1604 he married Elizabeth Browne, the daughter of a prominent London physician. They had no children and little is known of his wife, although they seem to have been happy together. Harvey refers to 'my dear loving deceased wife' in his will. In 1607 he was elected a fellow of the Royal College of Physicians and continued to be active in their affairs for the rest of his life. In 1618 he was appointed physician to James I, and continued to be physician to Charles I on his accession to the throne. Harvey was short, dark eyed and, in his youth, dark haired with a short beard. It is said that he was quick tempered (much like his brothers), that he carried a dagger, as was the fashion for well-to-do men at the time, and was quick to draw it when offended. There is also a story that Harvey, having done some experiments and dissections in public to demonstrate the circulation, threw down his scalpel and stalked out of the room in exasperation when one of his opponents, Caspar Hoffman, refused to be convinced by what he had seen. Harvey was a devout Christian and, like many others in the seventeenth century, saw no tension between his scientific work and his religious beliefs. Towards the end of his life he said that:

The examination of bodies has always been my delight, and I have thought that we might thence not only obtain an insight into the lighter mysteries of nature, but there perceive a kind of image or reflection of the omnipotent Creator himself.

Harvey's most important and most famous work is *Exercitatio Anatomica de Motu Cordis et Sanguinis in Animalibus*, which was published in 1628, with an English version in 1653. The title translates as 'Anatomical Exercises on the Motion of the Heart and Blood in Animals'. Sometimes this work is known as the 'Circulation of the Blood', but this is a little misleading. Harvey's work on the motion of the heart was an important (and equally innovatory) complement to his work on the motion of the blood, and it was important to Harvey that his conclusions could be applied quite generally to all animals, not just humans. Most commonly this work is referred to as *De Motu Cordis* and I will refer to it as *DMC*. As we shall see, *DMC* caused a considerable controversy, though Harvey's views were generally accepted by the time of his death.

Harvey was an ardent Royalist and took the Royalist side against the Parliamentarians in the English Civil War of 1642–1646. His close association with Charles I curtailed his activities with the Royal College of Physicians to some extent. Harvey's political views may be judged from the dedication to the King of *DMC*:

Most serene King! The animal's heart is the

basis of its life, its chief member, the sun of its
microcosm; on the heart all its activity depends,
from the heart all its liveliness and strength arise.
Equally is the king the basis of his kingdoms, the
sun of his microcosm, the heart of the state; from
him all power arises and all grace stems.

Harvey was with the King on the Scottish campaigns of 1639, 1640, 1641, was in virtually constant attendance on the King during the period of the English Civil War and was even present at the battle of Edgehill in 1642. During these years Harvey was in Oxford with the King, and he surrendered himself to the Scots in May 1646 at the close of the war. He attended the King in his captivity in Newcastle, and returned to London in 1647. Harvey seems to have got on well with Charles I, who took an interest in his scientific work. Charles allowed Harvey to dissect deer from his own herds, which was important for Harvey's later work on animal reproduction. Harvey also demonstrated experiments for Charles. Harvey must have been quite a remarkable person, dissecting and vivisecting animals frequently in the course of his research, but also managing to keep up the requirements of a gentleman and of physician to the King.

Harvey was busy with research, even during the war years in Oxford, but he published nothing further until 1649. We have a few of Harvey's letters from this period, but there were two major tragedies from the historian's point of view. Many of Harvey's papers, manuscripts and notes were lost when Parliamentary troops ransacked his house in Whitehall in 1642. In particular, he complained that he lost his book on the generation of insects, which contained the fruits of many years' research. He considered this loss to be the 'greatest crucifying' that he had in his life. Many of the rest of his notes were lost in the great fire of London in 1666, along with his new library, when the Royal College of Physicians' building was destroyed.

In 1649 Harvey published *Exercitationes duae de Circulatione Sanguinis* (Two Exercises on the Circulation of the Blood), in response to criticism of the circulation theory by Jean Riolan. Harvey's last publication was in 1651, when *Exercitationes de Generatione Animalium* (Exercises on the Generation of Animals, *DGA* for short) was printed. In his later years, he went quite grey and was troubled by gout, kidney stones and insomnia. He believed that he thought best in the dark, and liked to drink a good deal of strong coffee, which was just becoming fashionable in England (and may explain his insomnia!). He eventually died of a stroke on 3 June 1657, at the age of 79. Rumours that he took or attempted to take his own life with an overdose of laudanum, which he took for his gout and kidney stones, were vigorously denied by his friends and those who were present at his death.

De Motu Cordis

DMC was published in 1628, but when Harvey actually discovered the circulation of the blood has been a matter of debate. It is highly likely, though not absolutely certain, that he made the discovery at some point between 1617 and 1619. The evidence for this is that in the dedication for *DMC*, Harvey says that his views have now been confirmed by experiment for 'nine years and more', and that this book was 'otherways perfect some years ago'. In the preface, Harvey refers to Fabricius' 1615 book *De Respiratione* (On Respiration) as 'recently set out'. In Chapter XIII he refers to Fabricius as 'a venerable old man'. As Fabricius died on 21 May 1619, it would seem *DMC* must be from before then. There is also some evidence that Harvey's views were known in continental Europe as early as 1622. Evidence from his lecture note-book indicates he was not thinking about the circulation in 1616. So the most reasonable hypothesis is that Harvey had the idea of the circulation between 1617 and 1619 and wrote at least substantial and important parts of *DMC* around that time. In all likelihood he added the dedication later prior to publication, or updated it. It was not unknown for scientists, even in the seventeenth century, to try to backdate their discoveries, in order to have priority over their rivals. Harvey, however, unlike Newton on gravity and on some of his mathematical discoveries, had no significant rivals in being first to the circulation theory. In the sixteenth and seventeenth centuries there are cases where scientists were cautious of publication. One reason for this was fear of the church, but there was nothing controversial concerning religion in *DMC*. Another reason for delay in publication was often fear of derision, especially from fellow professionals, typically where a theory was radical and likely to upset long-held theories. Harvey was well aware of the revolutionary implications of his discovery, and in *DMC* VIII said that his ideas:

are so new and unheard of, that not only I fear

mischief which may arrive to me from the envy of some persons, but I likewise doubt that every man almost will be my enemy, so much does custom and doctrine once received and deeply rooted prevail with everyone.

Perhaps it is no great surprise, then, that Harvey sought to prepare the ground and support his ideas with a great many experiments before he made his theories more widely known. The structure of *DMC* is relatively straightforward. Harvey began by reviewing previous opinions on the motion of the heart and the blood, and moved on to his own ideas on the motion of the heart. Chapters VIII and IX are the pivotal chapters, as here Harvey argued for the circulation of the blood and described his most famous experiment, measuring the amount of blood passing through the heart. He then moved on to a series of supporting experiments.

The Circulation

How, though, did Harvey arrive at the theory of the circulation? Harvey's best argument for the circulation of the blood, and certainly his best known, is that there is so much blood flowing through the heart that the blood must circulate. The body cannot possibly produce and consume that much blood. The veins would rapidly empty and arteries rapidly distend if the blood did not find its way from the arteries to the veins and back to the heart. We need to distinguish between his best evidence, which may have been generated after he had the theory, and what led him to the theory in the first place. Robert Boyle (1627–1691) stated:

And I remember that when I asked our famous Harvey, in the only discourse I had with him, (which was but a while before he died), what were the things which induced him to think of the circulation of the Blood? He answered me, that when he took notice that the valves in the veins of so many several parts of the body, were so placed that they gave free passage of the blood towards the heart, but opposed the passage of the venal blood the other way: he was invited to imagine, that so provident a cause as nature had not plac'd so many valves without design: and no design seemed more probable, than that, since the blood could not well, because of the interposing valves, be sent by the [veins] to the limbs, it should be sent through the arteries, and return through the veins, whose valves did not oppose its course that way.

Harvey himself said that:

Truly, when I had often and seriously considered with my self, what great abundance there was, both by the dissection of things, for experiments sake, and opening of the arteries, and many ways of searching, and from the symetrie, and magnitude of the ventricles of the heart, and of the vessels which go into it, and go out from it, (since nature does nothing in vain, did not allot that greatness proportionably to no purpose, to those vessels) as likewise from the continued and careful artifice of the doors and fibres, and the rest of the fabric and from many other things; and when I had a long time considered with my self how great abundance of blood was passed through, and in how short time that transmission was done, whether or no the juice of the nourishment which we receive could furnish this or no: at last I perceived that the veins should be quite emptied, and the arteries on the other side be burst with too much infusion of blood, unless the blood did pass back again by some way out

of the veins into the arteries, and return in to the
right ventricle of the heart.

It seems that that the venous valves were very important in the process that led to the discovery of the circulation.

The Venous Valves

These venous valves are small flaps or cusps on the inside of the veins. They allow blood to flow in one direction, when the blood tends to push them against the vein wall. However, if the blood should attempt to flow in the other direction, the flaps move off the vein wall, 'catch' the blood and prevent it from passing. Should these valves fail to work, then blood can collect in a vein, and it can become varicose. The motions of the limb muscles help to push venous blood back towards the heart in concert with these valves. This is why it is good advice for anyone standing still for a great length of time (such as a soldier on parade) to wiggle their toes and flex their calf muscles.

Harvey showed that the venous valves all have a cardiocentric orientation, so that it would seem that their function was to prevent blood in the veins from flowing away from the heart. The old view was that the valves were gravity orientated, to stop blood congregating in the lower parts of the body. However, Harvey, who credited Fabricius with the discovery of the venous valves, told us in *DMC* XII that:

> *The finder out of these portals did not understand the use of them, nor others who have said the blood by its weight should fall downward: for there are in the jugular vein those that look downwards and do hinder the flow of blood upwards.*

Fabricius called the venous valves *ostiola*, 'little doors', and published a work on them, *De Venarum Ostiola*, in 1603. Actually, the venous valves were known to Vesalius, and even before him. As Vesalius thought these valves merely strengthened the veins, no one took a great deal of notice of this discovery. As for the nature and function of the venous valves, in *DMC* XII Harvey stated:

> *I have often tried in dissection if beginning at the roots of the veins I did put in the probe towards the small branches with all the skill I could, that*

it could not be driven by reason of the portals:
On the contrary, if I did put it in outwardly from
the branches towards the root, it passed very
easily.

Harvey was also at pains to point out that there are
no valves in the arteries. Venous valves are found in
many species of animal, but not artery valves. This
is something Harvey recognised in his early lectures
on anatomy, where he gives his own opinion that:
'There are many valves in the veins opposed to the
heart; the arteries have none except at the exit from
the heart'. That Harvey had recognised the nature
of the venous valves so early is good reason to
suppose he had the circulation thesis early as well.

The Ligature Experiments

Harvey performed some important experiments
using ligatures (tourniquets) in order to demonstrate that the blood flows out via the arteries,
passes in some manner to the veins, and then flows
back to the heart again. As Harvey put it in
DMC XI:

The blood doth enter into every member through

the arteries, and does return by the veins, and that the arteries are the vessels carrying the blood from the heart, and the veins are the vessels and wayes by which the blood returns to the heart itself; and that the blood in the members and extremities does passe from the arteries into the veins.

These experiments exploit the fact that in the arm the arteries lie deeper than the veins, so using a moderately tight ligature one can stop blood passing through the veins, but allow it to pass through the arteries. Using a fully tight ligature one can stop blood passing through both veins and arteries. With the first sort of ligature the veins will swell but a pulse will still be felt, with the second type there will be neither swelling nor pulse.

If we begin with a tight ligature, we find that the hand becomes cold but remains properly coloured. If we slacken the ligature somewhat, then the hand becomes flushed and the veins stand out and become distended. If we then release the ligature altogether, we release the blood in the veins and the arm returns to normal. Harvey concluded from this that blood was carried out in the arteries. With a tight ligature the arteries are cut off as well as the veins, so the arm does not swell and there is no pulse. With the lighter ligature the arteries still carry blood out, but as the veins are restricted they cannot carry blood back and so there is a swelling of the veins. So blood is carried out by the arteries, finds its way into the veins, and is carried by the veins back to the heart.

Harvey had a further set of experiments with ligatures, this time to demonstrate the direction of the blood flow in veins, and these experiments rely on the fact that the valves in the veins will allow the blood to flow in one direction only, towards the heart. Here we have one diagram from Harvey's *DMC* (Figure 10).

Firstly, we set up a ligature such that the veins become distended. There will also be small, localised swellings in the veins themselves. These will be either where veins meet (E, F) or where there are venous valves (C, D). If, using a thumb or a finger, one draws the blood in a vein down towards the hand (i.e. away from the heart), the blood will not pass the next venous valve (H). The vein above the finger will disappear as it will have been emptied of blood, but below the finger it will become very distended as far as the next valve. Keeping your finger on at H, try the same thing with another

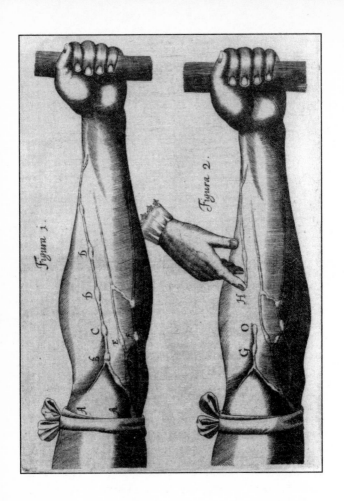

Figure 10 (and facing page): The ligature diagrams from William Harvey's *Exercitatio Anatomica de Motu Cordis et Sanguinis in Animalibus* (1628). The diagrams are almost identical to one in Fabricius' *De Venarum Ostiolis* of 1603. (Source: Wellcome Library, London)

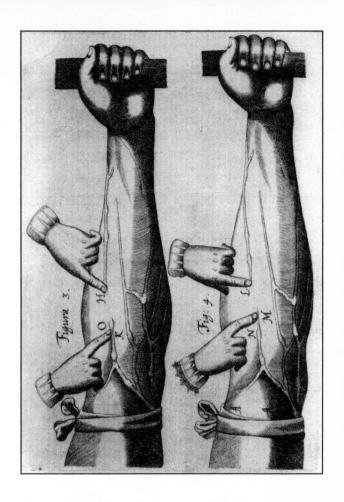

finger above the next venous valve along the arm, and you will get the same result. The part of the vein emptied by the first finger does not refill, but that blood moved by the second finger is stopped by the venous valve. The more effort you make to force the blood past the valve, the greater the swelling of the vein, but no blood will go past the valve. It would appear that blood can flow only in one direction in the veins.

Harvey described a second experiment on the same theme. Set up the ligature on an arm such that the veins become distended and small, localised swellings due to vein junctions and venous valves can be seen. This time, press one finger on a vein between venous valves and use the other to push blood up the arm towards the shoulder (i.e. towards the heart). Blood will pass the venous valve you are pushing it towards, but when you remove your finger the blood will not return past the valve. Release your first finger, which is below the emptied section of vein, and the vein will refill immediately. As the vein can only fill from below, and not from above, it appears that blood only flows one way in the veins, and that is towards the heart. Harvey confirmed these results with other experiments. He told us that:

In the exposed jugular vein of a doe (in the presence of many nobles and the most serene King, my master), divided in two across its

*length, scarcely more than a few drops of blood
came out from the lower portion, rising up from
the clavicle.*

Harvey did many experiments on deer from the
King's herd, and it seems, as here, that the King and
other nobles took an interest in Harvey's work.
Here the point is that although a major vein has
been severed, it does not bleed from the heart side
of that vein because the venous valves prevent
blood flowing away from the heart. Harvey continued: 'On the other hand, though the other
opening of the vein is a fairly long way down from
the head, a round column of blood came out very
copiously in a great rush.' The body side of the
vein bleeds profusely, as the venous valves will
not prevent flow in the direction of the heart.
Harvey also tied off the arteries and veins of living
animals in order to see which side became
distended with blood and which side became
denuded, thus indicating the direction of flow and
also giving a rough idea of the speed of flow. Harvey
finished by saying:

*You will be able to make the same observation
daily during the outflow of blood in phlebotomy*

[blood taking]. *For, if you press on the vein with a finger a little below the opening, the outflow of blood is satisfactorily arrested, but on release of the pressure, it flows out again in abundance as before.*

The Flow-Rate Experiment

Harvey gave what has frequently been described as a 'quantitative' argument for the circulation of the blood. This is that the blood is passing through the heart so quickly that the blood must circulate. The body could not possibly produce new blood to replenish blood passed through the heart so quickly and so, unless the blood circulates, the veins would empty rapidly. Galen's account of the two blood systems simply cannot work if so much blood is passing through the heart.

Harvey's calculation worked like this. Firstly, he gave three estimates of the volume of the left ventricle (3, 2 and 1½ ounces) and commented that in a cadaver he found it holding more than 2 ounces. Harvey used apothecaries' weights, where 20 grains = 1 scruple, 3 scruples = 1 dram, 8 drams = 1 ounce, 12 ounces = 1 pound. 1 scruple = 1.3 grams. He then gave five possible estimates of what proportion of this volume is ejected at each beat (⅓, ¼, ⅕, ⅙, ⅛). The actual value was around 50 per cent, or slightly less, so the relaxed volume of the ventricle was about twice that of the contracted volume. He then gave an estimate of the number of beats per half-hour for the human heart: '1,000, and in some cases 2,000, 3,000 or even 4,000.' These pulse rates were 33, 67, 100 and 133 beats per minute: 33 is rather low for a healthy human, 67 would be a reasonable resting pulse, while 100 and 133 are reasonable exercising pulses.

Harvey then concluded that, using 1,000 beats per half-hour and 0.5 ounces as the amount passed per beat, at least 500 ounces of blood would be passed. So, overall, the calculation works like this:

Volume of left ventricle × proportion of blood passed = amount passed per beat.

Amount passed per beat × beats per half-hour = amount passed in half an hour.

So, using the conservative estimates of 2 ounces for the left ventricle volume, ¼ for the proportion passed and 1,000 for the number of beats in half an hour we have:

$2 \times 0.25 = 0.5$ ounces of blood per beat.

0.5 × 1,000 = 500 ounces of blood per half-hour.

In modern units, this works out to 16 kg of blood per half-hour, roughly 0.5 kg per minute. Modern values here, typically, might be around 120 ml for the volume of the ventricle, around 50 per cent for the proportion expelled and around 70 b.p.m. for the heart rate. In fact the amount expelled can vary, being greater during exercise. These figures would give us 60 ml of blood expelled per stroke, and so 60 × 70 = 4,200 ml per minute, around 4 litres. Per half-hour, this will be 30 × 4,200 = 126,000 ml, 126 litres. As the body normally contains around 5 litres of blood, one can see the force of Harvey's argument in favour of the circulation of the blood, either using Harvey's figures or the modern ones. During exercise, a human can increase their cardiac output to 20–25 litres per minute, and a very fit person may manage 35 litres. Harvey did similar calculations for animals and came to the critical conclusion that: 'It is manifest that more blood is continually transmitted through the heart than either the food which we receive can furnish or is possible to be contained in the veins.' This is undoubtedly very good evidence that the blood must circulate. The body cannot possibly produce or consume so much blood as is being passed by the heart, even using the conservative estimates that Harvey employed. Mammals may produce relatively large amounts of milk when they are suckling young, said Harvey, but such amounts were passed by the heart in one hour. These underestimates add to the force of his argument, as accurate values will give figures that favour Harvey even more.

Harvey also performed an experiment to demonstrate that blood cannot pass from one side of the heart to the other, through the septum, as Galen and his supporters had believed. The pulmonary artery, pulmonary vein and the aorta of a cadaver are tied off, and the left ventricle opened. A tube attached to an ox bladder is introduced through the vena cava to the right ventricle. Warm water is then injected 'with great force' from the ox bladder, with the effect that the right ventricle and its neighbouring auricle 'swelled violently', and despite the fact that 'almost a pound' of water had been forced into the right side of the heart, 'not even a small drop' of water or blood escaped into the left ventricle.

The Motion of the Heart

Harvey also undertook a further very important

piece of work, which was to work out precisely what happens when the heart beats. This is by no means easy, as the beat happens very quickly and it is not at all clear to the naked eye what is going on. Harvey was quite aware of this, and said in *DMC* V that:

Nor is it otherwise done, than when in engines, one wheel moving another, they seem all to move together; and in the lock of a piece, by the drawing of the spring, the flint falls, strikes the steel, fires the powder, enters the touch hole, discharges, the balls flie out, pierces the mark and all these motions by reason of the swiftness of them, appear in the twinkling of an eye.

The difficulty of this is witnessed by the fact that many careful anatomists got this quite wrong previous to Harvey. The problem lay in working out which of three functions of the heart happened together.

- The heart is a muscle, so it has a cycle of contraction, relaxation and rest.
- The heart takes blood in from the veins and expels it into the arteries.
- The visible motion of the heart is that it rises and shortens and then sinks and lengthens.

Here we need to introduce two technical terms. Systole is the term used for the contraction of the heart and diastole is the term used for its relaxation. These terms are also used in the measurement of blood pressure. Blood pressures are given as systolic blood pressure/diastolic blood pressure, typically in a healthy human, 120/70 ('one twenty over seventy'). The first figure is a measure of the blood pressure when the heart contracts and sends a new expulsion of blood into the body, so raising the pressure in the arteries, and is known as the systolic blood pressure. The second figure is a measure of the pressure in the arteries when the heart is relaxing, and is known as the diastolic blood pressure.

It had been thought previous to Harvey that when the heart rises and shortens, the muscles are relaxing and the heart is filling with blood, and so the heart is in diastole. Harvey demonstrated that just the opposite is the case. When the heart rises and shortens, the muscles are in contraction, the volume of the atria and ventricles diminishes, and blood is expelled from the heart. When the heart rises and shortens it is in systole.

One part of Harvey's work here is detailed observation of hearts that beat slowly in order to see the precise sequence of motions more clearly. So Harvey looked at the hearts of cold-blooded animals, which in general beat more slowly than those of hot-blooded animals, and he also observed hearts as they die, when again they beat more slowly. As Harvey says in *DMC* II:

This is more evident in the hearts of toads, serpents, frogs, house-snails, shrimps, crevises and all manner of little fishes. For it shews itself more manifestly in the hearts of hotter bodies, as of dogs, swine, if you observe attentively till the heart begins to die, and move faintly, and life is as it were departing from it.

Harvey also observed that when the heart rises and shortens, it feels harder to the touch, as with muscles in contraction. It also becomes whiter when it rises and shortens, and redder once more as it drops and lengthens, as with muscles contracting and relaxing. Finally, Harvey pierced the ventricle and saw blood leap out.

Harvey was also careful in his analysis of the progress of the heartbeat. He noted that the atria contract before the ventricles, something that had been noticed before, but is more marked in slow-beating hearts. It is the closing of the heart valves, as first the atria and then the ventricles contract, that gives the heart beat its characteristic 'lub dup' sound, as it is described in medical texts. Harvey was able to produce an important conclusion from this about the flow of blood into the heart. He vivisected animals and watched for the heart to begin to stop beating. First the heart slows, then the ventricles stop contracting. With only the atria contracting, Harvey pierced the auricles and showed that they too expel blood. This is important evidence against those Galenists who thought that the heart in some manner attracted blood into it. The atria in fact push blood into the ventricles.

These observations of the heart, and the conclusions that Harvey drew from them, prompted him to look at the nature of the heart as a muscle much more closely. Vesalius had believed that the muscle fibres of the heart ran straight up and down, but Harvey showed that the arrangement of these fibres is more complex in order to be able to generate a more complex motion of the heart.

The Pulse and the Heart Valves

This new analysis of the heartbeat also meant that

systole now occurred at the same time as the pulse was felt in the arteries. Thus Harvey could declare, contrary to the opinion of Galen, that: 'The pulse of the arteries is nothing but the impulsion of blood into the arteries.' Galen and his followers had believed that the arteries pulsed of their own accord, as their account of the heart had systole and arterial pulse out of phase. Harvey also contradicted Galen's account of the function of the heart valves.

The two valves between the atria and the ventricles, the tricuspid and mitral (or bicuspid) valves, on the right and the left of the heart respectively, prevent blood from flowing back from the ventricles to the atria (see Figure 11). The two valves between the ventricles and the major blood vessels are the pulmonary valve and the aortic valve, which prevent blood from flowing back from the arteries into the ventricles. Galen believed, quite rightly, that the pulmonary, aortic and tricuspid valve all make good seals and do not allow any backflow of blood through them. However, he believed that the mitral valve allowed some passage of blood in what we would consider to be the wrong direction. Galen's theoretical need for this was that as the pneuma in the vivified blood was consumed by the body, the blood became fouled with sooty wastes

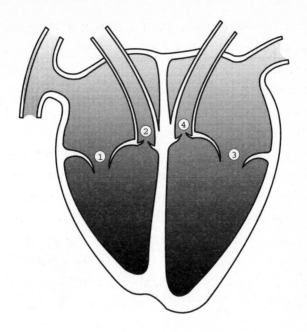

Figure 11: Diagram showing the heart valves.
Key:
1   Tricuspid valve
2   Pulmonary valve
3   Bicuspid (mitral) valve
4   Aortic valve

which could only be removed, and the blood re-
vivified, by the lungs. Thus blood had to be able to
reflux through the heart to the lungs. There was
also a need for nutritive blood, which had passed
through the septum, to become vivified and so this
had to be able to reach the lungs as well. He argued

that as the mitral valve has only two cusps, while the tricuspid valve has three, there must be some reason for this. In fact the heart valves do not leak in healthy patients. When valves do leak they create heart 'murmurs', the sound of turbulence as blood leaks through the valve, different from the normal clean 'lub dup' sound of the heart. Harvey, with the correct theory of the flow of the blood, was able to arrive at the correct theory of the functioning of the heart valves as well.

Harvey used one type of argument that may seem somewhat strange to us, but was nevertheless important to him, both in arriving at the circulation theory and in the experiments he used to demonstrate it. This was the principle that nature does nothing in vain. According to this, nature does not produce useless things, nor does it produce an excess of something where it is not necessary. So every part of the body has a function, and is designed to fulfil that function optimally. Nowadays, we would reject this principle on two sorts of grounds. Firstly, there are parts of the human body that have little or no function, such as the appendix. Secondly, the theory of evolution has nature doing a great deal in vain, if many variations on a species are produced of which only the fittest survive. That nature does nothing in vain was a standard idea for the seventeenth century, however. Harvey applied this principle to the heart and its major blood vessels. Why, he asked, if it only carries nutrition to the lungs, and a small proportion of the blood flow, is the pulmonary artery so large, indeed as large as the pulmonary vein? Surely the lungs do not require this much nourishment? Why is the left ventricle as large as the right? From their symmetry, Harvey argued, we should expect their functions to be similar. So, too, with the heart valves. Why is it, when the venous valves are competent at stopping backflow, and so are the aortic, tricuspid and pulmonary heart valves, that the mitral valve should be incompetent? We have also seen this principle in relation to the venous valves. There is a purpose to their design and orientation, and that is the cardiocentric flow of blood in the veins.

Harvey's discovery had two complementary parts. There was the idea of the circulation of the blood, and Harvey most likely arrived at this idea from his contemplation of the venous valves, and then supported it with the other experiments we have seen. Complementary to this remarkable and revolutionary idea was his analysis of the motion of the heart and its valves, which in its way was equally innovative and revolutionary. This also radically broke with Galen and other previous anatomists, and was to be as much a source of controversy in the seventeenth century as the circulation theory.

3   The Nature of Harvey's Discovery

So how modern was Harvey? The older view is that Harvey discovered the circulation because he employed the new methods and ideals of the scientific revolution of the seventeenth century. In this sense, Harvey is thought to be a modern thinker, in tune with the positive trends that were so important for the scientific revolution. However, there are many aspects of what Harvey did and said that simply cannot be accounted for in this way. The newer view is that far from being a modern thinker in this sense, Harvey was essentially an Aristotelian. He did not reject outright the works of the ancients, but in the Renaissance manner still revered them and sought to develop their programme. To see this, we need to look more closely at the nature of the scientific revolution, so we can judge to what extent Harvey's work was in tune with it. The briefest and simplest characterisation of the scientific revolution is that it was a change from an Aristotelian view of the world to a Newtonian one.

The Scientific Revolution

The scientific revolution began around 1543. This year marked the publication of the theory of Nicolaus Copernicus (1473–1543) that the sun and not the earth is at the centre of the solar system, and that the earth and the planets in fact orbit the sun. This was directly contrary to the views of Aristotle and the scholastics. After the work of Galileo Galilei and Johannes Kepler to improve the Copernican theory and provide evidence to support it, it was a major factor in the rejection of the Aristotelian view of the cosmos. The year 1543 is convenient because it also marked the publication of Vesalius' *De Fabrica* and the beginning of the criticism of Galen.

What were the major areas of change for the scientific revolution? Certainly there was a change in attitude towards the works of the ancient Greeks. One might say that the Middle Ages saw an un-questioning dependence on the works of the ancients, the Renaissance saw a critical dependence on the ancients, while the scientific revolution saw the complete rejection of ancient authority. Important figures in the scientific revolution in this regard were René Descartes and Francis Bacon, both of whom rejected the ancients utterly and began again from scratch in order to have a proper basis for science and philosophy.

A change in attitude towards the value of mathematics in science is a critical feature of the scientific revolution. Key figures here are Galileo, Descartes and Newton. It is easy to take some of the advances of the scientific revolution for granted, as we now see mathematics as absolutely fundamental for science. Many relationships are expressed as mathematical formulae, and it is believed that mathematics provides a precise description of a great many aspects of the natural world. The scholastics, however, gave much greater prominence to qualities than to quantities in their philosophy of nature. They described the world around them in terms of what they directly perceived rather than with numbers. So they used the terms 'hot' or 'cold' where modern science would assign a numerical temperature (e.g. 20°C) and used colour terms where modern science would talk of wavelengths of light. They used comparatives such as hot, hotter, hottest, rather than establishing a scale of numbers such as our temperature scale. They did not believe it was proper, or indeed possible, to quantify the world. Partly this was due to a lack of technology, but scholastic philosophy was the most important consideration here. For the scholastics, the natural world was not the sort of place that could accurately or usefully be described in terms of quantities. One of the great successes of the seventeenth century lay in changing this view. Galileo insisted that the language of nature was that of geometry and mathematics, and conducted beautiful, accurate quantified experiments. This laid the basis for the enormously successful mathematical physics of Newton. Equations such as 'force equals mass times acceleration' ($F = ma$), fundamental to our understanding of how objects move, and the idea that the intensity of gravity depends on the mass of the objects involved, the gravitational constant and the square of the distance between the objects ($F = Gm_1m_2/d^2$), were the fruit of the idea that the world could be described precisely by mathematics.

Along with this came a move towards the mechanisation of nature and an atomic theory of matter, led by Descartes, Gassendi and Mersenne in Europe and Thomas Hobbes, Robert Boyle and Newton in Britain. Again, it is easy to take the advances of the scientific revolution for granted, if we ignore the views that were discarded in the seventeenth century. The scholastics did not believe that matter could be ultimately broken up into discrete, individual pieces as in the atomic theory. Rather, they thought in terms of qualities again. Four of these were primary, arranged as two pairs – hot and cold, wet and dry (see Figure 12).

The four elements of their matter theory had one from each pair of qualities. So earth was cold and

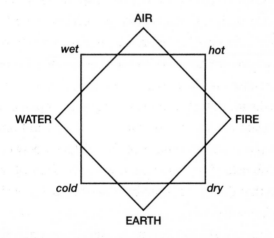

Figure 12: Aristotle's elements. Each element is characterised by a pair of the wet/dry and cold/hot opposites.

dry, water was cold and wet, air was hot and wet, and fire was hot and dry. The seventeenth century rejected this idea in favour of atoms that had no qualities as such, only quantities of length, breadth, depth, gravitational attraction and motion. These inert pieces of matter interacted in an entirely mechanical manner. We can see here the relation to mathematisation. The ultimate pieces of matter in the seventeenth century have only properties that are quantifiable and so are mathematically treatable.

In explaining the world, the scholastics emphasised the qualities, form and purpose of wholes and tended to use organic models (this works *like* a plant or an animal, not a machine). While this could work reasonably well for the human body, the scholastics tended to treat the inanimate objects of physics like this too. The seventeenth-century mechanical philosophy emphasised quantity and mechanical interaction, and tended to use mechanical models (this works *like* a machine, not something animate). So the human body was made up of small particles and functioned like a machine, and to use the favourite metaphor of the seventeenth century, like a very complex piece of clockwork. We have moved away from the crude initial application of these ideas (especially in biology), but the mechanical philosophy, with its atoms and mechanisms, was an important step forward for the seventeenth century.

Also important for the scientific revolution was the rise of experimental methods. The scholastics did not practice science as such, but there was for them a part of their theology/philosophy that dealt with the natural world. Their methods tended to be those of philosophy and theology, debate and disputation, rather than observation and experiment. The seventeenth century saw an important move towards experiment in investigating the natural world.

There were also important social changes for science. Science, rather than being merely that part of philosophy which dealt with nature, became a discipline in its own right, and became oriented to changing the world for the better rather than being a purely theoretical pursuit. Science (and scientists) changed in status, and there was a great deal more optimism towards what science and technology might be able to do. How did Harvey fit in with these changes? Did he reject the ancients and advocate mathematics and mechanism, or were there significant influences from Aristotle? One part of the case for Harvey as a modern thinker is that in creating the idea of the circulation of the blood, he broke radically with all ancient thought on the matter. However, this may be more complex than it appears at first sight. Harvey's teacher Fabricius did something characteristic of the Renaissance in attempting to revive the programme that he felt best typified the golden age of ancient Greece. This meant going back beyond Galen, to Aristotle. How did Aristotle's programme differ from Galen's? Aristotle's programme in anatomy was to investigate animals in general, rather than man in particular. So the idea was to find out what the function of the heart, or respiration, was in all animals. Answers had to be applicable to all animals and not just man. Fabricius' intention was not to replace Aristotle, but to revive the programme Aristotle began and take it further.

Harvey and Aristotle

Harvey had ample opportunity to learn from Aristotelian influences. He would have been exposed to a considerable amount of Aristotelian philosophy while he was at Cambridge. He may well have been influenced by the Aristotelian professor of philosophy Cesare Cremonini (1552–1631) while he was at Padua, and of course he was a student of Fabricius. In the last chapter, I insisted on giving *DMC* its full title – 'Anatomical Exercises on the Motion of the Heart and Blood in Animals'. That Harvey wished to consider the heart and the blood *in animals* is an important indication of the Aristotelian nature of his project. This is borne out in Harvey's work in dissecting and vivisecting animals in *DMC*. He sought a function for the heart, and a flow of the blood, that applied to all animals. This is important, for one criticism Harvey made of the Galenists was that they only ever considered the heart in relation to the lungs. Harvey was able to consider the function of the heart in animals without lungs, such as fishes, and so was able to come to more general conclusions. It is also interesting to note what Harvey said about his own research, and about Fabricius, in the first chapter of *DMC*: 'But so much the more willing I was to do it, because Fabricius having learnedly and accurately set down in a particular treatise, almost all the parts of living creatures, left the heart only untouched.' Harvey set out to investigate the heart, and in doing so discovered the circulation of the blood. To go back to the full title of *DMC* again,

Harvey was also interested in the motion of the *heart* and blood in animals. There is no indication in any of Harvey's works that he ever rejected either an Aristotelian programme in anatomy or a generally Aristotelian notion of the natural world. Indeed, in many places he praised Aristotle on his method. Certainly he did not reject Aristotle in the manner of Bacon or Descartes. Harvey's place in history is a complex one. He made a radical discovery, but did so as part of the revival of an ancient programme. There is considerable further evidence of a dependence on Aristotelian ideas.

Harvey and Quantification

The fact that Harvey conducted experiments to measure the quantity of blood flowing through the heart, and then produced a mathematical argument for the circulation of the blood, has been taken as one of the strongest points in favour of Harvey the modern. If we look at the nature of the experiments and the argument more carefully, however, a different picture emerges. The flow-rate experiments did not require, and were not given, any significant precision. All that was needed was a recognition that a much greater amount of blood was being driven through the heart in a given time than the veins could contain or the ingested food could supply. The values for the contents of the left ventricle that Harvey gave were only approximate and show no signs of any systematic quantitative survey to give minimum, maximum and average values. The values for the proportions transmitted were conjecture, and were presented as such, with no empirical support. The pulse rates given were clearly approximations, and indeed the value used in the computations for humans (1,000 per half-hour = 33 b.p.m.) was a significant underestimate. The same value was used for sheep, dogs and oxen, but was only a reasonable estimate for oxen.

It is clear that the figures Harvey used were not based on precise measurement. The number of species that Harvey had vivisected and dissected indicates that he was a diligent observer with a systematic approach. Having opened this number of animals, he could quite easily have produced comprehensive figures for the volume of blood in the ventricle, and indeed for other quantitative aspects of the circulation. Yet Harvey had no interest in quantification or quantitative experiment for its own sake. Here we might compare him with Galileo, someone whose use of quantified experiment and mathematics was critical for the seventeenth century. Galileo did beautiful experiments on how objects fall, made careful, systematic and accurate observations and derived mathematical laws about the motions of objects. Harvey merely gave as much in the way of quantitative information as his argument required and no more. Some of that is conjecture and the rest is estimation, and it is not evidence of a systematic quantitative method.

There is an interesting passage in Aristotle's *Meteorologica* where he argued against the idea that rivers are supplied with water by great underground reservoirs which fill up in the winter and then gradually deplete during the summer. He then argued that:

It is clear that, if anyone should wish to make the calculation of the amount of water flowing in a day and picture the reservoir, he will see that it would have to be as great as the size of the earth or not fall far short of it to receive all the water flowing in a year.

The nature of this argument was identical to that of Harvey's. The amount of liquid flowing (water/ blood) was too great for the opposed hypothesis (reservoirs/Galen) to account for. It is interesting that Harvey referred to the vena cava as the 'headspring of the veins, and cistern or cellar of the blood' and argued that the veins will soon empty if the blood did not circulate, given the flow of the blood through the heart.

Harvey's flow-rate experiment was merit-worthy and important, and was revolutionary in its implications for anatomy and physiology. It did not break new ground, however, in investigating quantities and in using a mathematical argument. Both Galen and Erasistratus used quantitative physiological arguments in antiquity, both being concerned with quantities of food and water ingested and the subsequent weight of the body and excretions. As Harvey knew the *Meteorologica* well, it is quite possible that reading Aristotle inspired this experiment. Neither this experiment nor anything else that Harvey said required any conception of the world as a place that was amenable to precise mathematical description. Nor is there any indication that Harvey would have been in sympathy with Galileo's highly influential idea that the book of nature is written in the language of geometry. Indeed, in the introduction to *DGA*, Harvey commented that '*Nature's Book* is so *open*, and *legible*', but made no reference to mathematics or geometry. While Harvey did make methodological comments both here and elsewhere concerning diligent and systematic investigation, he never commented on the use of mathematics and quantification. Harvey's use of quantification in the flow-rate experiment does not show that he was in tune with the developments of the seventeenth century.

Harvey and Mechanical Models

The second strong argument in favour of Harvey the modern is his supposed use of mechanical analogies for the heart. Harvey likened the heart, it is said, to a pump, and this is a classical mechanical analogy in line with the progressive trends of the seventeenth century. Again though, if we look carefully at what Harvey had to say, a somewhat different picture emerges. Harvey did not in fact liken the heart to a pump, but to a pair of water bellows, and did so only in his lecture notes. He said: 'From the structure of the heart it is clear that the blood is constantly carried through the lungs in to the aorta as by two clacks of a water bellows to raise water.' A water bellows is not an orthodox pump as its body is collapsible, while the cylinder and piston of an orthodox pump are rigid. So too the clacks were pieces of leather placed over the entrance and exit of the main chamber to allow one way flow only (see Figure 13).

Like the valves of the heart these are flexible, rather than the rigid valves found in orthodox pumps. This is significantly less of a mechanical analogy than might be suggested by a direct analogy with a pump. Aristotle, in his *De Respiratione*, likened not only the lungs but also the heart to a pair of forge bellows: 'It is necessary to regard the structure of this organ [the lung] as very similar to the sort of bellows used in a forge, for both lung and heart take this form.'

Galen also likened the heart to a forge bellows.

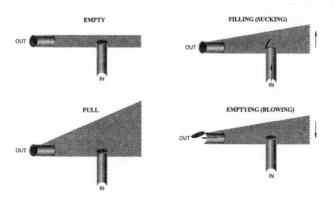

Figure 13: Diagram showing the function of clacks in a forge bellows.

While Aristotle of course did not have a mechanical account of the heart, he was quite happy to use this analogy with a forge bellows to get across the basic structure and functioning of the heart. Although Harvey's analogy was slightly more sophisticated, there was nothing which required a mechanical interpretation or differentiated Harvey's account from that of Aristotle in any essential matter. There is also a world of difference between the following:

(i) Saying while teaching (as in Harvey's lecture notes) that the heart can be likened to a pair of water bellows, to allow students to understand the structure and function of the heart.

(ii) Saying that the heart is in fact a pump.

Harvey and the Natural Magic Tradition

There is a very important passage in *DMC* that further indicates how deeply Harvey was influenced by Aristotle, and by some other interesting sources as well. Harvey argued that we may call the motion of the blood circular:

> *After the same manner that Aristotle sayes that the rain and the air do imitate the motion of the superior bodies* [sun, moon, planets and stars].

For the earth being wet, evaporates by the heat of the Sun, and the vapours being rais'd aloft are condens'd and descend in showers, and wet the ground, and by this means here are generated, likewise, tempests, and the beginnings of meteors, from the circular motion of the Sun, and his approach and removal.

So in all likelihood it comes to pass in the body, that all the parts are nourished, cherished, and quickened with blood, which is warm, perfect, vaporous, full of spirit, and, that I may so say, alimentative; in the parts the blood is refrigerated, coagulated, and made as it were barren, from thence it returns to the heart, as to the fountain or dwelling house of the body, to recover its perfection, and there again by naturall heat, powerfull and vehement, it is melted and is dispens'd again through the body from thence, being fraught with spirits, as with balsam, and that all the things do depend upon the motional pulsation of the heart.

So the heart is the beginning of life, the Sun of the Microcosm, as proportionably the Sun deserves to be call'd the heart of the world, by whose vertue, and pulsation, the blood is mov'd, perfected, made vegetable, and is defended from

corruption and mattering; and this familiar
household-god doth his duty to the whole body,
by nourishing, cherishing, and vegetating, being
the foundation of life, and author of all.

Here we see a classic example of a macrocosm/
microcosm analogy. The basic idea is that humans,
the microcosm ('little world'), have the same sort of
characteristics as, or function in a similar way to
something much larger, the macrocosm ('large
world'), like the earth, the solar system or the
universe. The first macrocosm/microcosm analogies were due to Plato (427–347 BC), one of the
great Greek philosophers, who argued that the
universe as a whole had a soul, life and intelligence
just as men have a soul, life and intelligence. Plato's
views underwent a revival in the Renaissance, and
macrocosm/microcosm analogies became quite
common in certain circles.

In the previous chapter, we saw another of
Harvey's macrocosm/microcosm analogies, likening
the King to the sun. As the sun rules and is the focus
of the heavens, so the King rules and is the focus of
his kingdom. Here we have the idea that the
weather cycle as described by Aristotle is the
macrocosm and the heart and the circulation of the

blood is the microcosm. The sun's heat evaporates water, which is then condensed, falls as rain, passes into streams and rivers and is then evaporated again (the macrocosmic cycle). Similarly, the heart warms and pumps blood in humans, which passes through the arteries, nourishes and is cooled by the body and then returns to the heart (the micro-cosmic cycle). Figure 14 is taken from Sachs A. Lowenheimb's work *Oceanus macro-microcosmicus* of 1664 and shows what Sachs supposed to be the macrocosm/microcosm relationship between the earth's weather cycle and the human circulation.

Harvey got something very important out of this macrocosm/microcosm analogy. He procured a means of answering two of the major difficulties for the circulation thesis which I mentioned in Chapter 1, the problem of the constant interconversion of arterial and venous blood and the rapid passage of large quantities of blood through the lungs and the body. Harvey recognised that there are two types of blood, venous ('rawish, unprofitable, and now made unfit for nutrition') and arterial ('digested, perfect and alimentative'). As these two types are now within one system, Harvey had to make their constant interconversion plausible. Moreover, he had to do so without knowing the modern account

Figure 14: Frontispiece diagram from *Oceanus macro-microcosmicus* (1664) by Sachs A. Lowenheimb.

of what is involved in these conversions. Here he leant heavily on his macrocosm/microcosm analogy. The weather cycle for Aristotle has the qualitative and cyclical changes (in Aristotelian terms) of water changing into air by evaporation and air changing into water by condensation. The analogy with Aristotle's weather cycle is very tight. As the sun provides heat for the macrocosm, so does the heart for the microcosm. That is all the more significant because the sun's heat generates the key change in the weather cycle, the evaporation of water. In Aristotelian terms, that is the change from cold/wet water to hot/wet air. The heart also effects a key change in the circulation in converting one type of blood into the other, and does so by its 'powerfull and vehement' natural heat. In Aristotle the contrary conversion is a cooling, as it is with Harvey, and the heart 'melts' the blood while the parts coagulate it. Finally, in Aristotle the sun is the cause of all change in the terrestrial realm. Without it, the elements would settle out into four concentric rings. For Harvey all things depend on the motion of the heart, and in the preface addressed to the King, it is that on which all growth depends and from which all strength and vigour proceeds.

Harvey also had the difficulty of discerning what

happens between arteries and veins in the absence of evidence from direct observation of the capillaries. So too there is a problem about the passage of blood through the lungs, each problem being made acute by Harvey's estimation of the quantity of blood flowing through the heart. Aristotle had an analogous difficulty, in that while rivers, evaporation and rainfall may be evident, it is less clear how rainfall becomes rivers. Aristotle hypothesises that the mountains act like a sponge, and that gradually water collects together and emerges as rivulets which then form the rivers. In *DMC* Chapter VII, where Harvey was talking of the passage of the blood through the 'streyner of the lungs', his leading example was that: 'It is well enough known that this may be, and that there is nothing which can hinder, if we consider which way the water, passing through the substance of the earth doth procreate Rivulets and Fountains.' Only then did he give the examples of sweat passing through the skin and urine through the kidneys. The latter are weaker examples, as they will not support a great enough volume of passing liquid. So again Harvey relied heavily on the macrocosm/microcosm analogy between the circulation and the weather cycle.

Harvey's dependence on Aristotle's conception

of the weather cycle is clear, but he also depended on the macrocosm/microcosm analogy, which is something we do not find in Aristotle. The macrocosm/microcosm analogy is something typical of the Renaissance natural magic tradition and the revival of Plato's thought in Renaissance Neoplatonism. This is not to say that there was anything mystical about Harvey's thinking. The natural magic tradition, while it believed there to be unseen sympathies between objects, believed these sympathies to act in law-like manner (in the same circumstances, the same thing would happen). Magnetism is a good example. Without any visible means of doing so, magnets could affect certain other objects at a distance (those they had some 'sympathy' with), and always did so in the same manner. Natural magic claimed to work in harmony with the natural powers of the world, and sought out hidden sympathies. In this way it needs to be distinguished from spiritual or demonic magic, which required the intervention of a spirit or a demon, and so required something (literally) supernatural or some action contrary to nature.

Some believed that there was a sympathetic link between macrocosm and microcosm, such that changes in the macrocosm might find corresponding changes in the microcosm. This formed one basis for astrology, where the motions of the planets and stars might find some correlating change in humans. Others believed that the world was created by a benevolent God in such a way as that there were considerable similarities between macrocosm and microcosm, and so we could use this analogy to explain something about humans or the universe.

Both natural and spiritual magic flourished during the Renaissance, and both effectively came to an end with the scientific revolution. The scientific revolution established a conception of the world as atomistic, mechanical and mathematical. It was a world with no room for the sympathies or spirits that the natural and spiritual magic traditions depended upon. The enormous success of the scientific revolution in many fields led to the general acceptance of this new view of the world. This new view could explain magnetism and many other phenomena important in natural magic in a straightforward mechanical manner without any need for recourse to hidden sympathies. Natural magic withered and died.

Harvey, however, lived before this new world picture was established. Prior to the scientific revolution, natural magic was a plausible competitor and complement to orthodox science. Harvey's use of the macrocosm/microcosm analogy should be no great surprise, nor should it cause us to think any the less of his work. Harvey employed what was, at the time, a perfectly reasonable way of looking at the world in order to solve some of the problems that the circulation thesis faced.

Alchemical ideas may also have been significant for Harvey. Robert Fludd was a doctor with a considerable interest in natural magic. He was a close friend of Harvey's and the first person to support Harvey in print. Fludd was explicit about an alchemical interpretation of the circulation. The process of heating, cooling and perfecting is seen as an alchemical process similar to that of the distillations of the alchemists. Fludd was keen on a broad interpretation of alchemy as something involving far more than the transmutation of lead into gold. Rather, alchemy was about perfecting substances in general, lead into gold being one example. If we look then to Harvey, it is interesting to look at how he wrote about the circulation. The blood is heated and in general perfected and made useful, alimentative and fit for nutrition by the heart, while in the body it is cooled, refrigerated, coagulated and made barren. The distinction between the two types of blood was that venous blood is rawish, unprofitable and unfit for nutrition, while arterial blood is digested and perfect. The word that Harvey used in the Latin version of *DMC* for the circulation, *circulatio*, was commonly used by alchemists for the process of distillation. In his lectures on anatomy, Harvey compared the functioning of the lung to that of the alembic, a favourite piece of apparatus among alchemists. Certainly the alchemical interpretation of Harvey was significant in the acceptance of the circulation thesis in some quarters.

Another important idea of natural magic and Neoplatonism was the primacy of the circle as a shape. There were many in the ancient world and in the Renaissance who believed the simple, elegant and unitary nature of the circle, and the fact that circular motion could continue without any changes, showed that the circle was the best shape. Giordano Bruno (1548–1600) was the one person prior to Harvey who believed the blood moved rapidly in a circle. He believed this because the soul, being intelligent, would choose the best motion, and so would move in a circle. We see this in the heavens and with the weather cycle of the earth, so we should also see it in the microcosm of the human body. As Bruno associated the soul with the blood, thus he reckoned the blood to move in a circle. As he had no other support for this idea, we rightly recognise Harvey as the true discoverer of the circulation. Harvey also used the analogy of the circle and its central point in relation to the heart. In his lectures on anatomy he said that: 'The heart ... is the principle part [of the body] for it occupies the principle place at the centre of a circle.' Circles and circular motion, that is motion that is constant and cyclical, held a fascination for many people in the sixteenth and early seventeenth century as the best shape and the best sort of motion.

Harvey and Witchcraft

Although Harvey was undoubtedly influenced by the ideas of the natural magic tradition, he had no interest in spiritual magic, mysticism or witchcraft. Harvey lived in the time of the European witch-hunt, and James I was a believer in witches and active in the persecution of alleged witches. Charles I was less active in this matter than James, but persecutions still took place. Harvey was involved in one of these, in 1634, and was called on to examine four women accused of witchcraft. Witches were supposed to have a hidden insensitive spot on their bodies, where the devil had touched them. This might take the form of a mole, a callus or any patch of skin that was insensitive when pricked. Witches might also have some form of third nipple for suckling their familiar. A familiar was a spirit or imp that took animal form. We associate witches with cats but, in the seventeenth century, rodents, frogs and birds were associated with witches as well. Harvey found nothing of note with three of the women, but his treatment of the fourth is interesting. He found that she had a piece of skin which formed 'a teat like that of a bitch' which was 'between her secrets', but that this was nothing that would not be expected from piles (haemorrhoids) or the application of leeches. He also found that she had a spot that was 'like the nipple or teat of a woman's breast', but that it was the same colour as the rest of the skin, without any hollowness and did not produce any blood or other juice. These four women were duly acquitted. Had Harvey been a believer in witchcraft, he could easily have construed what he had seen on the body of the fourth woman as the signs of a witch. People were convicted and hanged on far less evidence than Harvey found, but Harvey chose a different and rather more scientific explanation of that evidence.

The following tale turned up in a nineteenth-century manuscript. It may well be true, but we have no means of verifying it. A woman who was reputed to be a witch lived alone in an isolated house on the edge of Newmarket Heath. Harvey, with the King's leave, set out to discover if she really was a witch. Harvey was in Newmarket with the King in 1636, and women living alone, especially those keeping pets which might be seen as familiars, were particularly vulnerable to accusations of witchcraft. Harvey went to her house and pretended to be a wizard, to gain her confidence. The women then called her pet toad from under a chest, by making a clucking sound, and the toad drank some milk offered to it. Harvey got the woman to go to the nearest ale house for some beer, so they could drink together as brother and sister in witchcraft. While she was out he caught and dissected the toad. Harvey's dissection found nothing in the least unusual. Its anatomy was that of a normal toad (Harvey had dissected many in his researches). Its stomach was full of milk, showing it had really and not just apparently drunk the milk (familiars were supposed to have strange powers). The woman, not surprisingly, attacked Harvey when she found out what he had done. Harvey revealed that he was the King's physician, who had come to find out if she was a witch or not, and to arrest her if she was. This, and an offer of money in compensation for the toad, mollified her somewhat, though she still cursed him freely.

Again, people were executed for witchcraft on far less evidence than this. A woman was hanged in Cambridge in 1645 on the sole basis that she kept a frog as a pet. It is greatly to Harvey's credit that he approached these cases with an open mind and was willing to investigate carefully at first-hand and consider scientific explanations of the evidence allegedly showing witchcraft. It would have been very easy, at a time when belief in witches was rife and at times hysterical, and when it was formally heretical to deny that there were such things as witches, to interpret any suspicious behaviour or mark on the body as positive evidence of witchcraft.

Harvey and the Scientific Revolution

The scientific revolution saw a significant change in the nature of the explanation of physical pheno-mena. In short, scholastic explanations in terms of purpose and some ultimate good, and natural magic explanations in terms of harmony, sympathy and correlation, were rejected. In their place came explanations based on discrete, inert pieces of matter (atoms) and their mathematical properties. The dominant way of conceiving of how the world worked changed from the organic to the mechanical.

Harvey simply did not fit with this movement. As we have just seen, he leant on explanations from the natural magic tradition to some extent with his use of the macrocosm/microcosm analogy. After his discussion of the circulation and the macrocosm/microcosm, Harvey said: 'But we shall speak more conveniently of these in the speculation of the final cause of this motion.' Final cause was Aristotelian terminology for an explanation in terms of purpose and some ultimate good. Harvey disagreed with Descartes about the nature of the heartbeat. The root cause of the disagreement was that for Descartes, the human body was a mechanism and so the heart and blood were thought to work in a mechanical manner. For Harvey, the blood was alive and the heart beat of its own accord. Harvey, quite simply, was not a mechanist about the human body. Harvey was also scathing about materialism in *DGA*:

They that argue thus, assigning only a material cause, deducing the cause of natural things from an involuntary and casual occurrence of the elements, or from the several dispositions or contriving of atoms, do not reach that which is chiefly concerned in the operations of nature, and in the generation and nutrition of animals, namely the divine agent, and God of nature, whose operations are guided with the highest artifice, providence, and wisdom, and do all tend to some certain end, and are all produced, for some certain good.

What Harvey rejected here was precisely what the new mechanical philosophy was all about. If we want his attitude to the sort of philosophy he supported and the sort he despised in the most blunt form, John Aubrey, a seventeenth-century biographer and collector of gossip, tells us. When he asked Harvey in 1651 what books he ought to read, Harvey 'Bid me go to the fountain head and read Aristotle, Cicero and Avicenna and did call the neoteriques [newcomers, i.e. believers in the new philosophies] shitt breeches.'

4 The Reception of Harvey's Discovery

How was Harvey's discovery received? At first, there was considerable opposition. Some people well placed in the medical establishment had built their careers on a belief in a Galenic conception of the body. They would have recognised the revolutionary nature of the theory that Harvey was putting forward. This was not just another minor amendment to Galen's anatomy, something that might easily be accommodated within Galen's general conception of anatomy. It was not merely, to put it crudely, a disagreement about the plumbing of the heart and of the blood vessels. If the blood circulated, then Galen's theory of nutrition was wrong and the theory of vital spirit from the lungs was wrong. If Harvey was right about the motion of the heart and arteries (that the pulse is solely due to the heart) then Galen was quite wrong about that too. More fundamentally, the principles of attraction that for Galen determined the motion of nutrition and blood about the body and was basic to his idea of the functioning of the body must be wrong if the blood was driven round by the heart. What is more, a good deal of the therapy based on blood letting, which had been debated in great detail on the basis of Galen's anatomy in the previous century, would be redundant. Letting blood simply could not have the effects claimed for it if the blood circulated. In addition, the relation of the blood to the other humours of Galenic theory would need to be rethought.

We must also be aware that what we nowadays might take to be convincing evidence might not have seemed so straightforward to someone in the early seventeenth century. The evidence for the circulation of the blood was not just a matter of what one saw, but also of interpretation and argument. What we find to be the best piece of evidence, the flow-rate test, did not seem all that significant to some, due to their attitude to the importance of quantities and mathematics.

Harvey himself certainly seems to have been worried about the sort of reception *DMC* might receive. We have seen that he was worried that every man would become his enemy as he went against tradition, and in the preface he said that his theory was: 'A new and unheard of opinion concerning the motion of the heart and the circulation of the blood'. Furthermore, in Chapter I of *DMC* he said that:

At last, moved partly by the requests of my friends, that all men might be partakers of my endeavours, and partly by the malice of some, who being displeased with what I said, and not understanding it aright, endeavoured to traduce [speak ill of] me publickly, I was forced to recommend these things to the press, that every man might of me, and of the thing itself, deliver his judgement freely.

John Aubrey related that:

I have heard him say, that after his book of the circulation of the Blood came out, that he fell mightily in his practice, and that 'twas believed by the vulgar that he was crack brained; and all the physicians were against his opinion, and envied him; many wrote against him. With much ado at last, in about 20 or 30 years time, it was received in all the universities in the world . . . he is the only man perhaps that ever lived to see his own doctrine established in his own lifetime.

Actually, from the outset Harvey did have some supporters, but it is a reasonable outline of what happened. The first published reaction to Harvey that we are aware of was a positive one, and came from Robert Fludd in 1629 in his book *Medicina Catholica* (Catholic Medicine, though that title is slightly misleading as the book explores mystical and sacred medicine). He fully accepted Harvey's work on the motion of the heart and the blood, but for his own reasons. Fludd's background was in the Renaissance natural magic tradition and he was particularly enamoured of the possible alchemical interpretations of the interconversion of arterial and venous blood, and the use of the macrocosm/ microcosm analogy. Another important point for Fludd, this time on religious grounds, was Harvey's belief that the blood was in some sense alive, which suited Fludd's religious views very well. Fludd sought to reform knowledge along magical and religious lines, and some aspects of Harvey's circulation thesis suited this project nicely. Naturally, Fludd was keen to emphasise these, as was Sachs A. Lowenheimb, whose application of the macrocosm/microcosm analogy we saw in the last chapter, and several others. Indeed, although there was criticism of Harvey's work on many grounds, the macrocosm/microcosm analogy he employed did not come under fire.

Opposition to Harvey

The first negative reaction in print was from James Primerose in 1630. Primerose was an ambitious young English doctor, who seems to have been keen to make a name for himself. His book was specifically aimed at Harvey and the arguments of *DMC*, and was published in London in 1630. Primerose's attack on Harvey was essentially a restatement of Galenic ideas with no new experiments or observations. There were some important criticisms of Harvey, however, and this book did herald a major debate about the motion of the heart and of the blood. Harvey did not publish any reply to his critics until 1649, when he brought forth two anatomical essays addressed to Jean Riolan. Riolan, who was professor of medicine at Paris, criticised Harvey's views in his *Encheiridium Anato- micum et Pathologicum* (Manual of Anatomy and Pathology) in 1648 and his *Opuscula Anatomica Nova* in 1649. Harvey and Riolan knew each other and Harvey had a high regard for Riolan as an anatomist. Riolan attempted to reconcile the theories of Galen with Harvey. There was a con- siderable amount of work published that attacked or purported to refute Harvey. In 1635, Emilius Parisanus, a doctor with an interest in philosophy and an important member of the Venetian College of Physicians, published an edition of *DMC* with supposed refutations of Harvey interspersed between paragraphs of the text. We also have a few of Harvey's letters, one in reply to criticism from Caspar Hoffman in 1636. Hoffman was professor of medicine at Altdorf in Germany, and as such held an important position. He could influence the next generation of students for or against Harvey. They met in 1636, but Hoffman was unconvinced by Harvey's arguments and demonstrations (this is the source of the tale that Harvey threw down his scal- pel and left in exasperation). There are two other letters of interest, to Schlegel (1651), a student of Hoffman but an ardent supporter of Harvey, and Morison (1652), a Royalist doctor who had fled to

Paris and who did not return to England until the Restoration. It is a great pity we do not have more of Harvey's letters and papers from this period, so that we have a more detailed record of how he viewed the criticism of his theories. Whether he considered any of these criticisms as bringing the circulation theory into doubt is unlikely. Later, in *DGA*, he was to write that: 'I perceive that the wonderful circulation of the blood, first found out by me, is consented to almost by all: that no man hath hitherto made any objection to it greatly worth a confutation.'

Harvey was criticised on many points and for many reasons. Let us begin with the flow-rate experiment. Hoffman stated that:

Primerose . . . denies that there is much blood, nay more he clearly asserts that there is but little, but it appears much on account of its swelling up, or, as I would say its ebullition. Taking the logistic [calculation] *of his adversary, he comes from ounces and drachms to grains.*

Harvey's estimation was that between ⅓ and ⅛ of between 3 and 1½ ounces are passed, and this gives a maximum of 1 ounce and a minimum of ³⁄₁₆ of an ounce (1½ drams). In apothecaries' weight,

20 grains = 1 scruple, 3 scruples = 1 dram, 8 drams = 1 ounce, so 1 ounce = 480 grains. This was a considerable difference of opinion, so much so that Primerose could support Galen's theory of two blood systems as the body might be able to produce and consume his estimate of the blood passing through the heart.

There were several issues that become entangled in the debate about the flow-rate experiments. For some, there was a question of whether measuring quantities of blood flowing could amount to a proof, especially as there were so many estimates involved. They recognised two sorts of proof. There was the sort of proof we see in geometry, where one might prove that the internal angles of a right angle always add up to 180, and the sort of proof where it could be shown that something could not be otherwise. Measuring or estimating quantities of blood and then making calculations did not fit into either of the categories and so the flow-rate experiments and calculations were of little consequence to them. Thus Hoffman accused Harvey of 'abandoning anatomy for logistic', in making calculations about the heart and the blood flow, and said that: 'Truly, Harvey, you are pursuing the incalculable, the inexplicable, the unknowable.'

The Theory of 'Ebullition'

While there were certainly those such as Hoffman who took this view, Primerose took a different line of approach. It was generally agreed (by Harvey as well) that there was some sort of heating effect as the blood passed through the heart. Primerose took the view that this heating effect caused the blood to swell up, 'ebullition' as Parisanus termed it. Some envisaged the blood frothing up when it passed through the heart, like boiling milk or honey, and greatly increasing its volume. If so, then a small amount of liquid blood might take up a considerable amount of volume when in this frothed up state. The ventricle would contain blood in this state. So the volume of a ventricle would not be a direct measure of amount of liquid blood it contains. If the frothing were considerable, the ventricle might only contain a small proportion of liquid blood. That might have a significant effect on the volume of blood that a ventricle passed with every beat. The consequence of this was that even if one accepted Harvey's conjectures on the size of the ventricle, the proportion of blood it expelled per beat and the beats per minute, the amount of blood that the ventricle expelled may be considerably less than Harvey supposed. So instead of the 'ounces and drachms' supposed by Harvey, the heart only passes 'grains', an amount that might easily be accommodated within Galen's theory. In short, Primerose's argument was that because of this ebullition, Harvey was wrong by orders of magnitude in his estimates for the flow-rate experiment.

Riolan, seeking to harmonise Harvey and Galen, proposed a much slower circulation around a compromise, restricted circuit. So according to Riolan, blood was generated in the liver and was carried by the veins, part of it going to the right side of the heart. Part of this blood then passed through the septum and became arterial blood, and was pumped slowly by the heart into the arteries. Some of this blood then passed slowly back into the veins, thus completing an attenuated circuit. This had the advantage for Riolan of maintaining the distinction between nutritive and vivified blood, which was central to Galen's theory. If blood passed through the heart at one or two drops per beat, as the ebullition theory suggested, then Riolan calculated that it circulated around the body once a day.

One effect of the discussion of the ebullition

theory was to re-open the debate about the motion of the heart and systole and diastole again. For Harvey, the upward motion of the heart was the systole (contraction) and this expelled blood. Those who accepted the ebullition theory changed this around. The upward motion of the heart was due to the rapid expansion of the blood, and so was the diastole. It was the downward motion that expelled the blood and so was systolic. This supported Galen's view of the motion of the heart (though not the reasons for it) and had the systolic expulsion of the blood and the pulse in the arteries at different times, so one could again argue along with Galen that the arteries pulsed of their own accord. Harvey replied to the ebullition theory in his second letter to Riolan. He stated that if equal amounts of venous and arterial blood were drawn from the same animal, they both clotted at the same time, clotted to the same consistency, expelled the same amount of serum, and ended up the same volume and colour when clotted and cold. Thus arterial blood was not more frothy than venous blood, or it would have reduced to a smaller volume as it became clotted and cold, as we would find with frothy milk or honey.

What is Actually Seen?

There was also the possibility that there were considerable differences between a live, functioning heart and any dead heart that might be seen during a dissection. It was one thing to measure the volume of the ventricle in a dead heart, as Harvey had done, but what of the live heart? Can we be sure that the volume of the ventricle of a dead heart is, or is even approximately, the volume of the ventricle of a living heart? And what effect does the fact that the heart is pushing blood into arteries that are already filled with blood have on the proportion of blood that a ventricle pushes into the arterial system at each beat? One of Primerose's criticisms of Harvey was that the estimates he made in relation to flow rate were not based on direct observation, but on conjecture. To the extent that Harvey had not made his observations and measurements on a living human heart, however much he had performed human dissections and animal vivisections, that was fair comment.

One of Harvey's supporters, Waleus, a prominent seventeenth-century anatomist who worked at Leyden, attempted to demonstrate the rapid circulation of the blood. He cut off the tip of the ventricle of a dog in vivisection, and copious amounts of blood were expelled, travelling up to 4 feet. He commented that 'So it is evident that blood is propelled by this part' and calculated that the blood circulated in around 15 minutes. Now, while this was undoubtedly spectacular as an experiment, it would have been unlikely to convince hardened opponents who would be well aware of how fast blood spurts from the heart and from the arteries and would have had an explanation for it. Waleus wrote a book in support of Harvey called *De Motu Chyli et Sanguinus* (On the Motions of the Chyle and the Blood) which was published in 1640.

There was, of course, a major difficulty for Harvey here in that he could not do experiments with a live human heart. The closest he came to any such experiment was with the unfortunate eldest son of Viscount Montgomery. While still a young boy he had suffered a fall which had broken the ribs on the left side of his chest. The wound had not healed properly, and had left a cavity through which the internal organs could be seen to some extent. In fact the heart could be seen and indeed touched. Harvey went to see him, and brought him back to the court of Charles I in order to demonstrate the motion of the heart and the nature of systole and diastole to the King. Harvey says that he observed that in diastole the heart was drawn in and retracted, and that when in systole it was thrust out and lifted up to the ribcage and beat against it.

Using this distinction between live and dead humans, there were those who wished to hold on to the idea that there were pores in the septum. They accepted that these pores could not be seen in dissection, and that no fluids could pass across the septum of a dead heart. They argued that this was because the pores collapse with the onset of death. Thomas Winston, who lectured at Gresham College from 1615 onwards, and was professor of Physic there, claimed that the pores in the septum could be seen in a boiled ox heart, an observation that found its way into several textbooks. Some claimed to be able to find pores in the human heart in careful dissection, others claimed there were no such pores and only by rupturing the septum could a probe be pushed through it. Harvey's answer to this, in his first letter to Schlegel, was the bladder and heart experiment described in Chapter 2, where water was forced at pressure into the right side of a dead human heart with all the vessels tied off. No water was found to pass through the septum.

Other Objections

There were also objections concerning the persistent interconversion of arterial and venous blood. There are, as we have seen, significant differences between arterial and venous blood, and indeed between the arteries and the veins, but in Harvey's time there was no understanding of oxygenated/deoxygenated blood. That arterial blood and venous blood are two states of the same basic blood seemed unlikely to some. Typical of these was Ole Worm, a professor at Copenhagen. In his 1643 letter to his nephew Thomas Bartholin, who was considering the problem of the heart and blood, he said:

Have we not demonstrably shown that the blood in the arteries differs from that in the veins – differs in substance, colour, subtlety and all other properties? Bethink you, I pray, whether the blood in the arteries can possibly be that of the veins?

Harvey, as we have seen, had the analogy of the weather cycle where the blood was heated by the heart and cooled by the body to explain the interconversion of arterial and venous blood. He also referred to the lungs as strainers, such that only the finer and brighter part of the blood passed through them. In all fairness to Harvey, this was rather vague and hopeful. It is no great surprise then that not everyone was immediately convinced concerning the constant interconversion of arterial and venous blood.

In his second letter to Riolan, Harvey suggested that there were three main reasons why some people believed that there was a fundamental difference between arterial and venous blood. Firstly, the colour of arterial blood was more florid. Secondly, in the dissection of cadavers, the left ventricle and all the arteries were found to be empty, while the veins were full. Thirdly, arterial blood had more spirit and so took up more room. Harvey answered the first point with the blood-clotting experiment described above. If venous and arterial blood were identical in colour once they cool and clot, there could not be any fundamental difference between them. Harvey denied that the arteries were always empty in death. According to him, this depended on the manner of death, and in those who had suffered death by drowning you would find the arteries and the veins equally full. Harvey was not impressed by casual references to spirits making the arterial blood take up more space. In his second letter to Riolan, he commented: 'So it is not surprising that these spirits, with their nature thus left in doubt, serve as a common subterfuge for ignorance.'

We saw in the last chapter that Harvey was interested in some sort of purpose for the circulation, some 'good' that the circulation serves. This caused him problems with his critics. They could see why the heart and blood vessels might perfect the blood, and that this was for the good, but not why the blood should become corrupted and then have to be perfected again. The standard pattern of such arguments in the seventeenth century was to help explain why things became good or more perfect, and did not include permanent corruption and perfection. So a child will grow to be an adult (the best and most perfect form of a human), but the adult does not become a child and then an adult again. When the adult corrupts, it grows old irreversibly. Corrupting what is perfect and then perfecting it again seemed to go against the principle that nature did nothing in vain. While this may all look rather strange to us, those who took this sort of argument seriously in the seventeenth century (and there were many who did) would have been distinctly worried that the blood, having been perfected, was made corrupt again by the body. So we can find Hoffman saying that:

You would seem to impose upon Nature the character of a most rude and idle artificer who destroys the work she has done and perfected so that, forsooth, she should not be at a loss for something to do, for, to make raw again the blood that has been perfected by all her agents and then again to concoct it, what is it other than this?

So too Primerose commented that:

But let it be so to your good pleasure, as some say, to what purpose is it done? To be concocted a second time. To be concocted? Does it not, therefore, become raw again? Unless this happens, how is it that bile is not created rather than blood? For all the physicians are of the opinion that in its final elaboration, blood is turned into bile, now yellow, now black. But let it be as you say, does not Nature concoct in order to concoct? And then afterwards concocts a second time? And repeats the process again a

*third time, and a fourth, and so on for ever? To
what purpose, I ask again?*

Harvey's answer to this in his second letter to
Riolan was: 'How many things are accepted in
physiology, pathology, and therapy of which we do
not know the causes, but of the existence of which
we have no doubts?' That looks a reasonable
answer, and at first sight even a modern answer. It
meant, however, that Harvey had to abandon some
of his comments on the purpose of the motion of
the heart and blood. That, in a seventeenth-century
context, where many were looking for some good
purpose for the circulation and for the interconversion of blood in particular, was something of
a drawback for the acceptance of Harvey's theory.
Some of his supporters were not so reticent in
finding a good purpose for the circulation. Waleus,
in his *De Motu Chyli et Sanguinus*, stated:

> *Blood circulates for the sake of its perfection.
> By virtue of its continuous movement it is
> attenuated. It warms up and becomes rarefied
> in the heart, and subsequently condensed and as
> it were more concentrated in the outer parts of
> the body. For none of its parts is warmer than the*

heart and none cooler than the surface. Hence a
kind of circulation operates, not unlike that by
means of which chemists utterly refine and
perfect their spirits.

There were also those who simply accused Harvey of attempting to destroy traditional anatomy and medicine. Harvey, in his first letter to Riolan, said that: 'The concept of a circuit of the blood does not destroy, but rather advances, traditional medicine.' In his second letter to Riolan, he said: 'To all these let my reply be that the facts manifest to the senses wait upon no views, the works of Nature upon no antiquity: for there is nothing older or of greater authority than nature.'

There was also the problem of the capillaries and the completion of the circuit of the blood. Harvey believed the capillaries to exist, and had produced evidence that he thought showed that blood must pass from the arteries to the veins. Hoffman had doubts about the capillaries, and said:

Here I would like to learn from Harvey by what
paths [the blood] goes out of the arteries into the
veins . . . let it be granted that invisible passages
exist, unknown to so many preceding centuries

and first seen by this Englishman, by what faculty is the backward movement made?

There was, at this stage, no direct observational evidence of the capillaries. So perhaps we should not be too surprised that Harvey did not convince all of his opponents straight away. Hoffman's further problem was with the return part of the circulation. Where we see the blood as being driven around its circuit, Galenic anatomy worked much more on principles of attraction. What attracts blood back to the heart is what worried Hoffman, though if we accept the full circulation and the full flow rate, this is not a problem.

History may treat Galen and his followers somewhat harshly on the subject of the pores in the septum. Galen could not see a flow of blood across the septum, or the pores in the septum, but such a hypothesis was necessary for his system. However, Harvey could not see the capillaries either (they are seen later in the seventeenth century with the microscope), but such a hypothesis was necessary for *his* system! Galen and later followers were sometimes derided for putting theory before observation. What difference was there, however, beyond Harvey being right and Galen wrong? Both had reasonable systems requiring us to believe in something unobservable.

It is interesting to note that Francis Bacon – the most encyclopaedic, and possibly the greatest, writer on science in his age – ignored the circulation of the blood, and this despite the fact that Harvey treated him. Harvey does not seem to have admired Bacon greatly, though he had studied his work. John Aubrey related that: '[Harvey] would not allow him to be a great philosopher. Said he to me: He writes philosophy like a Lord Chancellor, speaking in derision.' The problem between them may have been the question of Aristotle's philosophy. Bacon was a vitriolic critic of Aristotle, wishing to sweep away all he had said to make room for a new science. Harvey, wishing to develop an essentially Aristotelian programme of research, still had a great deal of respect for Aristotle.

Harvey's views generally had been accepted by the middle of the seventeenth century. The great English philosopher and scientist Thomas Hobbes wrote that Harvey was: 'The only man I know, who, conquering envy, hath established a new doctrine in his life-time.'

Henry Power, an early fellow of the Royal Society, said in 1652 that:

Amongst all the rabble of his antagonists, we see not one that attempts to fight him [Harvey] at his own weapon, that is by sensible and anatomical evictions to confute that which he has by sense and autopsy so vigorously confirmed.

Power is a little over-effusive here perhaps, but this comment is indicative of the fact that Harvey's views had won the day.

The Dispute with Descartes

I will deal with Harvey's dispute with Descartes separately, as it illustrates the gulf between Harvey's conception of the body and that of the mechanical philosophers whose ideas come to prominence in the middle of the seventeenth century. Harvey and Descartes agreed that the blood circulated around the body. Indeed, Descartes was very supportive on this matter, and he commented that: 'This circular motion of the blood was first noticed by an English physician called Harvey, who deserves the highest possible praise for making such a valuable discovery.' However, Descartes goes on to say:

But Harvey was not so successful, in my view, on the question of the heart's movement. He imagined, against the general opinion of medical men, and against the evidence of what we see, that when the heart lengthens, its cavities increase in size, and when it shortens, they become narrower. I claim, instead, to demonstrate that they become even larger when the heart shortens.

So while Descartes agreed with Harvey on the circulation of the blood, he had a different account of the motion of the heart. Descartes' view was a version of the ebullition theory. Blood entering the heart increased in volume and became frothy due to the innate heat of the heart. It was this that caused the heart to expand. Descartes described this innate heat of the heart as 'fire without flame', and as the blood swelled it was supposed to close the heart valves behind it. Here we have the old debate about systole and diastole again, this time with Descartes supporting the idea of the circulation of the blood, but with an ebullition theory of the heartbeat. Descartes believed that all of Harvey's evidence on the motion of the heart could be accommodated equally well by his ebullition theory.

This was because Descartes had a philosophical programme of his own to pursue. The mechanical philosophy treated the world in terms of discrete pieces of matter and the mechanical interactions between them. Descartes wished to extend this approach beyond the physical sciences, such as physics and chemistry, and use it for the human body as well. The human body, for Descartes, up to and including the brain (but not the soul), was entirely mechanical. Animals had no souls, for Descartes, and so were also entirely mechanical.

Descartes was quite happy to accept the idea of the circulation as this could be made to fit very well with his mechanical conception of the human body. What he could not accept was that the blood was alive, and the heart had a capacity of its own which enabled it to beat. Descartes needed an account of the heart whereby it could be seen as purely a mechanical entity. The theory of heat and frothing allowed him this. Harvey believed the blood to be in some way alive. In *DGA* we can find him saying that:

Seeing therefore that blood acts above the

powers of the elements and is endowed with such notable virtues and is also the instrument of the omnipotent Creator, no man can sufficiently extol its admirable and divine faculties. In it the soul first and principally resides, and that not the vegetative soul only, but the sensitive and the motive also.

Harvey was appalled at the mechanistic interpretation that Descartes gave his work! Harvey had his supporters in the debate about systole and diastole. Sir Kenelm Digby, an English free thinker with an interest in natural magic, agreed with Harvey in 1644 that the 'heart has a proper motion of its own', and attacked Descartes. Digby asked a rather awkward question for Descartes, and indeed anyone who held the ebullition thesis. Why is it that the hearts of some animals continued to beat even when they had been removed from the body? The heart of a snake, he found, would beat for up to 24 hours after it had been removed. This surely cannot be because of blood entering the heart. The control of the heartbeat is also a problem, if the heart is merely a mechanical entity. Surely the speed with which the heart beats is not solely due to the amount of blood available. Digby was joined in these criticisms by Vopiscus Plemp, who did not believe the heart to be hot enough to create a frothing effect. Plemp was notable as an important seventeenth-century theorist, who began by opposing Harvey but later became convinced by him. Descartes' reply, which is somewhat unconvincing, was that blood dripping from the atria of an excised heart was sufficient for it to carry on beating. Careful experiments by others showed that even when there was no blood visible at all in an excised heart, it kept beating. Descartes attempted to modify his position, but the ebullition theory was finally laid to rest.

Harvey's work on the motion of the heart and of the blood stirred up considerable controversy. He was up against highly intelligent people, well versed in anatomy and with a long and sophisticated tradition of Galenic anatomy and medicine to draw upon. We need to consider how things looked to such people. Even some of those who supported Harvey (such as Fludd and Descartes) had their own agendas and wished to emphasise some parts, but not others, of Harvey's work. The debate on the motion of the heart and blood was very wide-ranging. There were a great number of issues at stake. There was the motion of the heart, the pathway of the blood, the existence of pores in the septum, the flow rate through the heart, the teleology of the cardiovascular system, the existence of capillaries, the possibility of the ebullition of the blood. There were also debates about the importance of direct observation, what could be learned from dissection as opposed to vivisection and animals as opposed to man. Further implications of Harvey's discovery were that fundamental Galenic ideas about therapy and blood-letting theory, and the role of attraction in anatomy, needed to be completely rethought.

Many who opposed Harvey had an interest in the status quo, but we need to recognise that Harvey's evidence and arguments for the circulation thesis, while very good, were not immediately compelling to everyone in the context of the early seventeenth century. Because his conclusions were so radical and important, every part of the theory came under intense scrutiny. This in many ways adds to the magnitude of Harvey's achievement. When we recognise the strength of the opposing positions and the number of objections they could bring against Harvey, it is all the more remarkable that he managed the mental leap to the circulation thesis and the arguments and experiments which ultimately led to its acceptance.

Harvey's Later Work

Harvey's later work attempted to come to an adequate understanding of how animals reproduce and resulted in his second major work, *Exercitationes de Generatione Animalium* in 1651. In fact this had been a life-long project, and Harvey was able to do many experiments and dissections on the King's herd of deer. *DGA* was again strongly influenced by Aristotelian ideas, and to some extent by ideas from the Renaissance magical tradition as well (Harvey described the hen's egg as a microcosm). One thing that Harvey was keen to assert was the primacy of the blood. Not only was the blood alive, but Harvey was keen to show in his studies of embryology that it was the blood that formed first after an animal was conceived and that the organs formed later. He was keen to show, in particular, that the blood formed before the heart did. In *DGA*, Harvey made one interesting comment on the furore raised by *DMC*:

You know full well what great troubles my former lucubrations [writings] *raised. Better is it certainly at some times to grow wise at home, than by the hasty divulgation of such things, to*

the knowledge whereof you have attained with
easy labour, to stir up tempests that may deprive
you of your leisure and your quiet for the future.

Perhaps here we see the reasons why Harvey delayed the publication of DMC for some years after he had the idea of the circulation, and why he put little in print concerning the debate over the circulation until his letters to Riolan.

One of the odder tasks Harvey performed in his later years was the post-mortem of 'Old Parr', a man who was reputed to be 152. The Earl of Arundel heard of this old man living on his estates in Shropshire, and brought him to London. Thomas Parr then became something of a minor celebrity. First married at 80, he did penance for adultery at 105 and was married for a second time at 112, and had done farm work until he was 130. Fêted in London, when asked by the King about his religious beliefs, he replied that he found it safest to be of the same religion as the reigning monarch, as he knew that he 'came raw into the world', and accounted it 'no point of wisdom to be broiled out of it'. Unfortunately, he did not survive very long, his death being put down to a change to a rich diet and the unhealthy air of London after the country life of Shropshire. Harvey conducted the post-mortem with his usual diligence and acute observation, noting down how well each of the organs had fared in their supposed extreme old age.

Conclusion

Harvey's discovery did not come completely out of the blue. We saw the beginnings of criticism of the Galenic theory with Vesalius, and there was the subsequent revival of investigative dissection in the sixteenth century. The art and artists of the Renaissance undoubtedly aided the more accurate investigation of the human body. Several people (Ibn al-Nafis, Servetus, Colombo) had suggested the lesser circulation, and it is likely that Harvey knew of Colombo's work. Bruno had even suggested a rapid circulation of the blood, though on speculative grounds only, and it is unclear whether Harvey was aware of this. Fabricius' discovery of the venous valves was clearly important for Harvey, as was Colombo's work on the motion of the heart.

Important though this background is, Harvey managed something both innovative and revolutionary in proposing the full circulation of the blood.

His great achievement was to have the imagination and insight to conceptualise the circulation, against 1,500 years of Galenic tradition, and at the same time to produce the new theory of the motion of the heart which was necessary to make the circulation theory work. The discovery of the circulation of the blood goes hand in hand with a new theory of the functioning of the heart and arteries. Harvey also won the battle against the entrenched defenders of Galen. As I have stressed, these were intelligent, well-educated and experienced men, who put up a considerable rearguard action. This helps to show the magnitude of Harvey's achievement. He succeeded in the face of considerable opposition. No one else, however intelligent and well versed in anatomy, had even proposed a radically new theory in the past 1,500 years.

I began this book by outlining how remarkable it was to infer the workings of the cardiovascular system from the resources available in the seventeenth century. This is a point worth emphasising at the close as well. Harvey had no direct access to a functioning human cardiovascular system. Dissection could tell him a good deal about structure, but not about movement or function. The vivisection of animals could only provide a limited amount of information. Harvey was remarkable in marshalling what was already known and combining it with his own observations and ingenious experiments to create a convincing argument. The motion of the heart *is* rapid and complex. It is not at all evident from simple observation of a beating heart or dissection of a dead heart what the heart does or how it does it. There are questions about the function of the valves, how the volume of the heart changes and when the heart contracts and relaxes. Nor is it evident how quickly the blood flows or how, or even if the blood vessels link up in other parts of the body. Harvey not only had the ability to envisage the new theory of the circulation, but also had the ingenuity to create an extraordinary collection of experiments, observations and arguments to support it. There was detailed and careful anatomical work allied to ingenious and deeply thought out arguments, as with the venous valves and the flow-rate argument. There were ingenious non-invasive experiments such as the series of ligature experiments. There were the extraordinary experiments to slow down the heart, so as to see the way it beats more clearly. Harvey not only had great vision, but also was a great experimentalist and precise observer of nature as well.

The significance of Harvey's discovery went far beyond a reform of our conception of the cardio-vascular system. Galen was clearly wrong about the heart and blood, but this meant he was also wrong about digestion, nutrition, respiration and aspects of organ function. Galen's physiology depended to a large extent on principles of attraction. Harvey's work showed this was quite wrong for the heart and blood, as the heart forces blood around the body, and prompted a complete re-evaluation of the attraction principle. A significant part of medical therapy in Galen's system was based on blood letting. Blood was drawn from specific parts of the body, determined by Galen's conception of the blood system and the position of the organ thought to be affected, in order to treat a disease. If Galen was wrong on the blood system, however, this entire mode of treatment needed to be rethought. The deeper significance of Harvey's discovery then was that it helped instigate a complete rethink of the way that the body worked and how it should be treated. Ultimately, Galen's anatomy and physiology was not in need of improvement, as the Renaissance anatomists had attempted, but of complete rejection.

Our twenty-first century understanding of the

heart and blood is based on that of Harvey. Certainly, we understand some things in a more sophisticated manner, not least the role of oxygen, but the notion of a full and rapid circulation of the blood, with the heart forcing the blood around that circuit, remains unchanged. That view came about in the seventeenth century because of Harvey's ingenuity, imagination, perseverance and his remarkable use of experiment, observation and argument. Harvey did something quite remarkable. He produced a radically new theory of the cardio-vascular system, along with the supporting evidence and arguments, which overturned 1,500 years of thinking on the matter. That was a revolution in science in itself. This revolution led the way to a further revolution in thinking about the human body, and effectively laid the foundations for the modern study of anatomy and physiology.

Further Reading

There are several editions of *DMC*, usually with the two letters to Riolan added. The most readily available of these is the Everyman publication, preferably the 1990 edition with an introduction by Andrew Wear.

The latest (and very good) major work on Harvey is Roger French's *William Harvey's Natural Philosophy* (1994).

The two most important older works are Walter Pagel's *William Harvey's Biological Ideas* (1967), which is the seminal statement of the view that Harvey was heavily influenced by Aristotle, and Gweneth Whitteridge's *William Harvey and the Circulation of the Blood* (1971), which is the classic statement of the view that Harvey was a modern thinker.

Also interesting is Robert Frank's *Harvey and the Oxford Physiologists* (1980), which concentrates on the reception of Harvey's views.

For the radical view that Harvey discovered the circulation relatively late, and that *DMC* was written in two parts, see Jerome Bylebyl's *The Growth of Harvey's De Motu Cordis*, in the *Bulletin of the History of Medicine* (1973).

The standard biographies of Harvey are Geoffrey Keynes' *The Life of William Harvey* (1966) and Kenneth Keele's *William Harvey* (1965).

Charles Singer's *A Short History of Anatomy from the Greeks to Harvey* (1957) is now a little old, but is still a very useful introduction, and Nancy Siraisi's *Medieval and Early Renaissance Medicine* (1990) is excellent.

Two recent general histories of medicine are *The Cambridge Illustrated History of Medicine* (1996), edited by Roy Porter, and *The Greatest Benefit to Mankind* (1997), written by Roy Porter.

On the ancients, the best place to start is Geoffrey

Lloyd's *Early Greek Science: Thales to Aristotle* (1970), which has an excellent chapter on Hippocrates, and his *Greek Science After Aristotle* (1973) has chapters on Galen and on later Greek medicine.

Glossary

Alembic: A type of glass flask used in alchemy for distilling.

Alveoli: Small sacs in the lungs where deoxygenated blood exchanges carbon dioxide for oxygen and becomes oxygenated blood.

Anastomases: Small vessels cross-connecting larger blood vessels.

Anatomy: The study of the structure of the human body.

Aorta: Artery carrying blood from the left ventricle of the heart.

Aortic valve: Valve between the left ventricle and the aorta.

Arteries: Blood vessels carrying blood away from the heart.

Artery-like vein: Galen's terms for the pulmonary artery.

Atria (sing. atrium): The upper chambers of the heart, where blood enters the heart.

Auricles: Term used by Harvey for the atria. The upper chambers of the heart.

Bicuspid (mitral) valve: Valve between the left atrium and left ventricle.

Capillaries: Fine blood vessels connecting the arteries and the veins.

Cardiovascular: To do with the heart and the blood vessels.

Chyle: A Galenic term for food that has been digested but not yet turned into blood.

Clacks: Small pieces of leather nailed over the entrance and exit of a pump that function as valves.

Deoxygenated blood: Blood that is purple in colour, and is depleted in oxygen.

Diastole: Expansion, as when the heart receives blood from the veins.

Ebullition: A theory about the motion of the heart whereby the heart is thought to heat the blood, which then froths up like milk or honey being boiled, and so expands considerably in volume. It could be used to argue that the heart did not beat of its own making, or that only a small amount of blood was passed in each beat.

Lesser circulation (pulmonary transit): A theory held by Ibn al-Nafis, Servetus and Colombo that blood passes from the right ventricle via the lungs to the left atrium. Blood was still produced by the liver and consumed by the body, and so did not circulate around the body.

Ligature (tourniquet): A device used to constrict the flow of blood in a limb.

Macrocosm/microcosm: An analogy used in the Renaissance natural magic tradition and neoplatonism. The greater world (the macrocosm, typically the universe or the earth) is reckoned to have similarities with or function in the same way as the smaller world (the microcosm, typically humans), and vice versa. So the earth's weather cycle might be compared with the circulation of the blood in the human body.

Microcosm: See macrocosm/microcosm.

Neoplatonism: A revival of the thought of Plato in the Renaissance.

Nutrified blood: One of Galen's two types of blood. Generated in the liver, it passes hrough the veins carrying nutrition to the body. *See also* vivified blood.

Ostensor: In pre-Renaissance dissections, a man who pointed out the parts of the body being lectured about.

Oxygenated blood: Blood which is scarlet in colour, and is rich in oxygen.

Pericardium: A membrane around the heart.

Physiology: The study of the functioning of the human body.

Pneuma: A vital spirit thought by some ancients to be present in the air, thought by Galen to be transferred from the air to the arterial blood (making it 'vivified').

Pulmonary: To do with the lungs.

Pulmonary artery: Blood vessel leading from the right ventricle of the heart to the lungs.

Pulmonary transit (lesser circulation): A theory held by Ibn al-Nafis, Servetus and Colombo that blood passes from the right ventricle via the lungs to the left atrium. Blood is still produced by the liver and consumed by the body, and so does not circulate around the body.

Pulmonary valve: Valve between the right ventricle and the pulmonary artery.

Pulmonary vein: Blood vessel leading from the lungs to the left atrium of the heart.

Scholasticism: A Christianised version of Aristotle's philosophy, very important in the Renaissance and prior to the scientific revolution.

Septum: Muscular wall separating the left side of the heart from the right. Contrary to Galen's beliefs, it is impermeable to blood in healthy adults.

Systole: Contraction, when the heart expels blood into the arteries.

Tricuspid valve: Valve between the right atrium and right ventricle.

Vein-like artery: Galen's term for the pulmonary vein.

Venous valves: Small flaps on the internal walls of the veins that allow the blood to flow one way only, towards the heart.

Veins: Blood vessels carrying blood to the heart.

Vena cava: Vein carrying blood into the right atrium of the heart.

Ventricle: The lower chamber of the heart.

Vivified blood: One of Galen's two types of blood. It is vivified in the lungs and carries this life-giving principle via the left side of the heart and the arteries to the rest of the body. *See also* nutrified blood.